RESIDUALS MANAGEMENT IN INDUSTRY:

A Case Study of Petroleum Refining

CLIFFORD S. RUSSELL

PUBLISHED FOR RESOURCES FOR THE FUTURE, INC.
By The Johns Hopkins University Press
Baltimore and London

RESOURCES FOR THE FUTURE, INC.

1755 Massachusetts Avenue, N.W., Washington, D.C. 20036

Resources for the Future is a nonprofit corporation for research and
education in the development, conservation, and use of natural resources
and the improvement of the quality of the environment. It was established
in 1952 with the cooperation of the Ford Foundation. Part of the work
of Resources for the Future is carried out by its resident staff; part
is supported by grants to universities and other nonprofit organizations.
Unless otherwise stated, interpretations and conclusions in RFF publications
are those of the authors; the organization takes responsibility for the
selection of significant subjects for study, the competence of the
researchers, and their freedom of inquiry.

This book is one of RFF's industry studies, which are conducted under the
Quality of the Environment program and directed by Allen V. Kneese and
Blair T. Bower. It was edited by Tadd Fisher. Charts were drawn by Clare
and Frank Ford, and the index was prepared by Terry Pagos.

RFF editors: Mark Reinsberg, Vera W. Dodds, Nora E. Roots, Tadd Fisher.

Copyright © 1973 by The Johns Hopkins University Press
All rights reserved
Manufactured in the United States of America

The Johns Hopkins University Press, Baltimore, Maryland 21218
The Johns Hopkins University Press Ltd., London

Library of Congress Catalog Card Number 72-12367
ISBN 8018-1497-9

Library of Congress Cataloging in Publication data will be found on the last printed
page of this book.

FOREWORD

Clearly any rational effort to develop pollution control policy must concern itself, among other things, with the technologies available for reducing the discharge of deleterious substances to the environment and with the associated costs. Even the Federal Water Quality Act Amendments of 1972, which at first glance bypass such concerns by setting stringent national effluent standards, ultimately acknowledge considerations of technological and economic feasibility.

Nonetheless, the act is still quite narrow in its recognition of the implications of particular pollution control policies. The same is true of the legislation pertaining specifically to gaseous and solid residual materials. In principle the organization of the Environmental Protection Agency at the national level, and of analogous agencies in many states and localities, reflects recognition of the tight interrelatedness of environmental problems. At the same time, legislative efforts to control particular types of discharges of residual materials are still kept neatly in their respective boxes.

That this is a dubious procedure can be inferred from some quite general observations. The mass of materials and energy flowing into and through the production and consumption processes is conserved. This means that at a given level of these activities the overall mass of residuals cannot be reduced by a treatment process except as this may be part of a materials recovery or recycling activity. Actually, treatment as such must increase the total flow of residuals, since treatment itself requires inputs. This is not to say that treatment (or, more generally, residuals modification) may not produce less harmful substances than were originally generated—indeed that is its intent. Nevertheless, when policy makers view environmental media in separate boxes, it is quite possible that the pursuit of quality in one medium has the unintended effect of increasing burdens on another as the form of residuals is changed. Waste water treatment extracts sludges or solids that in turn constitute a frequently serious disposal problem. Sometimes this is solved by incinerating these materials and so transferring them to still another environmental medium. It is readily seen that similar considerations apply to the treatment or modification of gaseous residuals.

Moreover, public policy has emphasized end-of-pipe treatment processes; for instance, rapid tax write-offs are offered for particular types of treatment technologies. Such policy neglects a variety of other adjustments that can influence the amount and type of residual generated. These adjustments include governing the quality characteristics of the inputs and outputs and of the internal materials recovery and recycling processes. The inadequacy of public policy is one of the reasons many economists and others favor effluent charges or taxes over subsidies and direct regulations that tend to specify particular technologies. Charges or taxes provide an incentive to the individual residuals generator to choose an optimal set of residuals control technologies from among those available to him.

In his study, Clifford Russell provides a quantitative model that for the first time incorporates the full range of technological and cost considerations into a totally coherent mode of analysis. The model is implemented for plants in a major U.S. industry—petroleum refining. It conserves mass and permits the quality of inputs and outputs to be altered. It can recover and recycle materials, and it can treat residuals once generated. It optimizes control technologies for specified output mixes and residuals discharge limits and for gaseous, liquid, and solid residuals simultaneously. All this is done in terms of an objective function that considers the prices of conventional inputs and outputs as well as those associated with residuals control. By means of this model Russell is able to calculate the costs of various sets of residuals discharge restrictions differing in both stringency and combinations of substances. He also is able to test hypotheses about the importance of considering the industry's various residuals streams simultaneously and the significance of indirect factors like the quality characteristics of inputs and outputs and the prices of certain conventional inputs.

The methodological innovation of Russell's study invites attention. The years following World War II saw the perfection of a powerful new optimization technique—linear programming. Its first applications were military, but its applicability to industrial problems was soon exploited. Thus programming models of petroleum refineries and steel mills were built to aid managers in scheduling operations, optimizing output mixes, and anticipating the effects of price changes of inputs and outputs. These models helped to achieve the internal goals of the firms. But since the residuals disposal service provided by the environment was unpriced and for the most part otherwise unrestricted, the costs imposed on society by environmental degradation nowhere entered into the *private* profit and loss calculus of the firm. The model reported in this book is an extension of the previous type of programming model toward incorporating a *social* dimension into the industrial optimization problem. It does this by pricing or otherwise restricting the use of social (or common property) resources in the operation of the industrial plant. It furnishes a mechanism for inferring the impact of an internalization of previously external costs on the whole range of industrial design considerations.

The approach can be extended even further. Instead of restricting discharges directly, they can be fed into models of the natural environment and translated into various kinds of impacts on "receptors" throughout a "problem shed." Thus the impacts on receptors, a consideration more directly pertinent to policy, can be restricted rather than the residuals discharges themselves—although of course a residuals restriction is implied. Now, however, the approach is based on the calculated effect on a receptor rather than on an arbitrary restraint on the behavior of the industrial plant. Furthermore, multiple industrial plants and other residuals-generating activities can be fitted into such models of the environmental system, and optimization or other strategies played out for them can be taken all together. It is even possible to introduce political (or collective choice) considerations into such a regional residuals management model and to test how various alternative government forms in the region affect decisions on environmental quality.

This excellent study is the result of the imagination and diligence of the author. But it is also a product of an evolution in thinking at RFF about the industry's impact on environmental media. The first RFF industry study, published in 1966, was devoted solely to the intake and depletion of water in a major industry—electricity generation. Conducted by Paul Cootner and George Löf and published in 1966 under the title *Water Demand for Steam Electric Generation: An Economic Projection Model*, it reflected the specific concern of the time with potential shortages of resource inputs. The next study, *The Economics of Water Utiliza-*

tion in the Beet Sugar Industry (1968), by George Löf and Allen Kneese, reflected the broadened recognition that one of the main adverse impacts of industry on water supply—often, indeed, *the* main impact—is the quality-deteriorating effect of residuals discharges. Accordingly, the economics and technology of reducing waterborne residuals were its main concerns.

About the time the latter study was published, Blair Bower, who has given general direction to the RFF industry studies as well as participating in them himself, began to emphasize the interdependencies among the various residuals streams and the importance of the qualitative characteristics of inputs and outputs. RFF's later industry studies have progressively incorporated these considerations. Russell's is the first of the current set to be published, but other studies of the steel, pulp and paper, coal-energy, and canning industries are in various stages of completion.

January 1973   ALLEN V. KNEESE
                                Director
                                Quality of the Environment Program
                                Resources for the Future, Inc.

ACKNOWLEDGMENTS

I am heavily indebted to Blair T. Bower of Resources for the Future for initial inspiration and continuing encouragement. Those familiar with Bower's work will recognize this model as an outgrowth of his conceptual writing on industrial water use. The other person whose help has been absolutely essential is Elizabeth D. Mortland, director of RFF's computer group and the house expert on IBM's Mathematical Programming System. She put together and submitted every computational run done with three versions of this model.

Thanks are also due to the National Water Commission, which, through a research contract with RFF, supported the initial research and the construction of the first version of the model; to Allen Kneese, for encouraging improvement of this first version under the Environmental Quality Program at RFF and for helpful comments at subsequent stages; to Jack Dart, petroleum industry consultant, for early assistance in devising a flow chart; to Jeffrey Vaughan for diligent research in the refinery literature; to Bechtel Corporation for an extremely useful technical critique of the second version of the model; and to Dee Stell for superb typing at every stage of the work.

Others who have been kind enough to read and comment on earlier drafts of the manuscript include V. Kerry Smith, Milton Searl, Joseph Fisher, and Henry Jarrett (all of RFF); and W. F. Curran and E. Couper of the American Oil Company, Whiting, Indiana.

December 1972   CLIFFORD S. RUSSELL

CONTENTS

LIST OF TABLES

LIST OF FIGURES

I

INTRODUCTION

In the past decade, perhaps only in the past three or four years, public opinion in the United States has swung sharply away from the traditional view that the volume and blackness of smoke from factory chimneys are direct measures of community well-being. With that tendency to embrace the extremes for which we are so often criticized, we have reversed ourselves and have begun to demand that in the future all discharges of substances or energy into the natural environment be proscribed and that present discharges in excess of existing standards be regarded as criminal acts.[1] Many observers have pointed out that such flirtation with extremes leaves us open to severe disillusionment, and that it nearly always works against the achievement of our long-term best interest. In the case of environmental quality, there is considerable fear among those engaged professionally and politically with environmental issues that the bloom is even now off the rose. For example, they suspect that many promoters of the latest rash of seemingly tough new laws recognize that, while the laws appear to provide the final solution to the pollution problem, they are probably unenforceable.

Under present laws it is not difficult to see some potential sources of future disillusionment. First, despite occasional pronouncements attaching dollar values to benefits (present damages), there is only the sketchiest public knowledge of what we are likely to gain by controlling, or in-

[1] The water quality legislation that passed the Senate unanimously in the fall of 1971 provides in its present form that *all* discharges to the nation's waterways must end by 1985.

1

deed by eliminating, discharges of the leftovers of our production and consumption processes to the nation's air, water, and land resources. Expectations, fueled by the rhetoric of some environmentalists, may *already* have outrun our ability to deliver, if only in the sense that dramatic improvements are unlikely. For example, since it is so difficult to measure the effect of existing concentrations of sulfur dioxide (SO_2) oxidants and particulates on respiratory morbidity and mortality rates, any improvement resulting from cleaner air is probably going to be at least as hard to identify and in any event will not be such as to impress itself immediately and obviously on the public consciousness.[2]

A second source of disillusionment built into present law is a reliance on administrative orders and hearings and on the judicial process to obtain compliance with policy directives. This route has already demonstrated that it is productive of incredible delays, the cost of the best lawyers frequently being less than the cost of the required pollution control. Such delays are likely to sour the public on the entire operation, particularly since each new law is touted in terms of what it *would* accomplish *if* it were obeyed and if all the proposed subsidy funding were available.[3]

Third, the public has not been and is not being told in any meaningful terms what proposed and enacted legislation is likely to cost beyond the appropriations for subsidies to municipal treatment facilities.[4] This lack of sound information leaves a hole in the environmentalists' armor into which industry has already begun to drive a wedge or two.[5] In particular,

---

[2] It has been waggishly observed that the air pollution battle would be a good deal easier if we had songbird kills analogous to our fish kills. By extension there would be less danger of disillusionment if birds could begin to appear again in certain cities or neighborhoods after emission reductions the way fish have reappeared in the Thames at London. There is some hope in this direction; since coal burning was halted, Londoners have begun to observe birds unknown in the city for a century.

[3] The failure of appropriations to match authorizations in this area is, of course, a major reason for the lack of visible progress so far. But I am concerned here primarily with the features of the law that determine whether the nation is likely to feel it is getting what it is paying for, however small the payment.

[4] In this connection, aggregate projections of national costs over the next so many years, whether discounted or not, can certainly not be considered meaningful. Even if it were planned to finance all such costs through existing national taxes so that some rough notion of distribution were available, the numbers are simply too large to grasp. But, in fact, the costs will be distributed in many ways—some prices and taxes will rise, some profit rates and wages will fall—and no individual can have the faintest notion of how he is liable to be affected on balance by these many adjustments.

[5] The best-publicized example so far is the Union Carbide case. When the Environmental Protection Agency finally pressured Union Carbide to agree to take immediate and long-range measures to control SO_2 emissions from its plant in Marietta, Ohio,

by exaggerating the costs of various proposed policies, whether in terms of effects on consumer prices or unemployment, it is relatively easy to split the public on the basis of self-interest.

Two other potential sources of disillusionment involve the procedures of standard setting themselves. These generally involve setting limits both on emissions and on ambient concentrations of various substances. But those who set standards now know too little about the natural environment to be able to guarantee even the consistency of these dual sets of restrictions.[6] It is possible that in many regions the meeting of discharge limits will not result in the desired ambient air quality. Conversely, in other regions the emission limits may be much stricter than those necessary to meet the desired ambient standard. The former possibility clearly carries with it a greater potential for mischief, since it will create public disappointment and a new series of squabbles about tougher abatement measures. But it should not be beyond the ingenuity of those opposed to the present environmental quality program to exploit the latter result as well.

Finally, decisions about standards, both nationally and regionally, are still made separately for the major receiving media: air, water, and land. These confines encourage the use of analysis in which the laws of conservation of mass and energy are ignored so that substances removed from the effluent stream of direct interest are conveniently assumed to disappear.[7] Since, in fact, reducing the discharge of one substance to one re-

---

the company said in effect, "All right, but we will have to lay off over 600 men because we will have to shut down partially." This threat was later withdrawn. Similar threats have been made by ITT-Rayonier in connection with a cellulose pulp mill in Port Angeles, Washington, and by U.S. Steel in San Francisco. Some older plants will, of course, have to shut down because it will prove impossible or simply too expensive to reduce discharges by required amounts. What is needed is information that allows separation of bluff from reality.

[6] I leave aside the question of whether the current method is the appropriate one for making these important social choices. See, however, Edwin T. Haefele, "Environmental Quality as a Problem of Social Choice," in Allen Kneese and Blair Bower, eds., *Environmental Quality Analysis: Theory and Method in the Social Sciences* (Johns Hopkins University Press for RFF, 1972), pp. 281–332; and Clifford S. Russell, Walter O. Spofford, Jr., and Edwin T. Haefele, "Environmental Quality Management in Metropolitan Areas" (paper presented at the International Economic Association meeting on Urbanization and the Environment, Copenhagen, June 1972).

[7] Dramatic real-world examples are not difficult to find. New York City has apparently never caught up with the flood of solid waste it unleashed by requiring apartment house incinerators to be upgraded or to shut down. This result appears to have been unanticipated. See also "Blunders in Pollution Control: A Case Study," in *Air and Water News*, 23 August 1971, pp. 7–8, for the history of a small foundry's problems with captured particulates.

ceiving medium is very likely to result in increasing the discharge of another substance to the same or another medium, it is likely that our new standards will cause some startling side effects that will create additional frustration.[8]

Granting for the moment that these criticisms of current policy and procedure have some validity, what relevance do they have to a monograph supposedly concerned with petroleum refining? Surely I am not going to suggest new measures of benefits from higher environmental quality or propose compliance procedures superior to those I have criticized. No, indeed. But I am going to discuss a model type that has been designed to allow some improvements in the other problem areas: cost estimation; simultaneous consideration of air, water, and land (solid) problems; and consistency of ambient and emission standards. And it is desirable to begin by providing a relatively broad justification to encourage the reader to keep one eye on the potential utility of the model rather than to succumb entirely to the details of construction and results that may become tedious at times.

Before more is said about the background of the model and its potential utility, however, terminology should be clarified. Familiar terms like "solid waste" and "pollution control" appeared in the opening paragraphs, but the rest of this study includes references to "residuals" and "residuals management." This change may strike the reader as odd, even affected, but there are sound arguments in its favor. First, the word "residuals" is used here to draw attention to the fact that the leftovers from human production and consumption are at the heart of the problem. The discharges discussed here are not accidental or unanticipated, nor are they produced specially by malevolent enemies of the earth to kill fish or create smog. Rather, the production (or consumption) process might be thought of as inputs of mass and energy (as well as factor services) flowing into a box; that process will necessarily generate an outflow of an equal amount of energy and an equal amount of mass (except for a tiny conversion of mass to energy). One or more of the energy or mass outflows will constitute the product. The other flows generally will have zero prices in existing markets—or at least prices below the variable costs of production, transportation, and so on—and are the process residuals.

[8] Only a material recovery process allows a reduction in the total mass of discharges, but even in such a process the need for auxiliary inputs may mean that an overall reduction is not actually achieved. This point will be discussed further, particularly in chap. 2.

The firm's only concern is the disposal of these residuals at minimum cost.[9]

In consumption processes, of course, the output is the maintenance of human life, or the provision of "utility." Aside from the "investment" implied by a growing population, consumption residuals in the aggregate will equal inputs. And thus, for the economy as a whole, material and energy residuals will equal material and energy inputs to production, and only recycling will actually reduce the magnitude of the residuals requiring disposal.[10] (Assuming, of course, that the recycling processes themselves do not result in the generation of more residuals, on balance, than the use of virgin material.) Second, it is implicit in the term "residuals management" that without 100 percent recycling (as in a spaceship) and without dependence on solar energy, some residuals will always require disposal. Thus the rational choice of forms, amounts, location, and timing of discharges to the natural world is the problem that must be addressed.

Note that residuals are defined in reference to existing prices; there is nothing immutable about the mix of material and energy requiring disposal after any particular production process. A slight change in relative prices may make it worthwhile to recover a substance previously disposed of as a residual—and vice versa. For example, steel mills once discharged water after once-through use from direct-contact cooling of coke-oven gas. Now nearly all coking operations involve recovery of some of the entrained chemicals (phenols and ammonia) in the process of recirculating the cooling water. This change, of course, is not attributable to a single price change; rather it reflects changes in chemical prices, in equipment costs, and almost certainly in existing or anticipated pressure for higher effluent quality. On the other side of the coin, increasing affluence and changing technology in the glass industry have led to a switch from recycled to no-return glass containers.

[9] Subject to the recognition that whether material and energy outflows are products or residuals depends on price relationships, it can be said that residuals are "generated" in the production process. Generation is measured *subsequent* to such activities as material or by-product recovery that are justified for the firm purely on market grounds at going prices—without any influence from residuals management actions. Residuals discharges are measured where they enter the natural environment, as in stack gas at the top of the stack or process water at the outfall to the river. Hence discharges reflect any processes intervening after generation that are adopted because of residuals management pressures.

[10] This view of the problem has been most carefully and persuasively stated by Robert U. Ayres and Allen V. Kneese in "Production, Consumption and Externalities," *American Economic Review*, vol. 59, no. 3 (June 1969), pp. 282–97; and idem, *Economics and the Environment: A Materials Balance Approach* (RFF, 1970).

In addition, in order to be perfectly clear on the costs imposed on industry by the residuals management actions of government, it is necessary to distinguish carefully between those steps that would be taken at current prices in the absence of residuals management and those taken only in response to such actions. This distinction may frequently be extremely difficult to draw in the real world of change and uncertainty, but if it is neglected the door to misleading claims is opened.[11]

To return to the background of and justification for the type of model of an industrial operation described in this monograph—the discussion below will deal separately with the questions of (1) cost estimation and (2) regional modeling aimed, at the very least, at achieving consistency between ambient and emission standards. Within both areas it is emphasized that simultaneous consideration of airborne, waterborne, and solid residuals is correct in principle and of unknown but potentially great practical importance.

ESTIMATING THE COSTS OF GOVERNMENT RESIDUALS
MANAGEMENT ACTIONS

Currently such data as do exist on which to base challenges to prophets of industrial doom lie primarily in scattered studies that frequently have very limited distribution. In addition, few of these studies take into account more than one problem under more than one assumption about ancillary conditions, and the analysts fail to consider a range of alternative solutions.

Naturally no single available study has all these deficiencies. The publications on industries in the federal government series, *The Cost of Clean Water*, for example, are widely available and do include consideration of some alternative conditions in estimating the costs of effluent quality improvement.[12] Thus the volume on refineries gives estimates of the

---

[11] For example, steel mills recycle "home scrap" (the scrap generated in the mill) as a matter of course at current prices of scrap, ore, finished steel, etc. But advertisements from the steel industry are currently telling us that steel makers have been good citizens—recycling residuals—for years; presumably out of the goodness of their hearts.

[12] U.S. Department of the Interior, FWPCA, *The Cost of Clean Water*, vol. 3, *Industrial Waste Profiles*, nos. 1–10 (November 1967). The ten profiles have the following titles: No. 1, Blast Furnace and Steel Mills; No. 2, Motor Vehicles and Parts; No. 3, Paper Mills; No. 4, Textile Mill Products; No. 5, Petroleum Refining; No. 6, Canned and Frozen Fruits and Vegetables; No. 7, Leather, Tanning and Finishing; No. 8, Meat Products; No. 9, Dairies; No. 10, Plastic Materials and Resins.

costs of reducing the discharge of waterborne biological oxygen demand (BOD), phenol, and sulfide from newer, "typical," and older refineries that contain a wide range of different petroleum-processing units.[13] But only waterborne residuals are discussed, and hence no allowance can be made for the effects of simultaneous regulation of gaseous and particulate residuals discharges. The study also cannot be used to estimate the effects of changing the product mix or of changing product quality requirements.

The widely distributed industrial water use studies from Resources for the Future—one for beet sugar production and one for thermal electric power generation—share with *The Cost of Clean Water* series the disadvantage of being limited to only one or two waterborne residuals.[14] (Both the beet sugar and the thermal electric power industries have the simplifying advantage of producing homogeneous outputs.) The authors of these studies do examine a range of methods of changing the residuals generation and discharge of the industrial operations in question; in particular, process changes and treatment and recirculation alternatives. But the examination is relatively unsystematic in the sense that it would be difficult to use the studies to predict the impact of changes in relative input prices or input qualities.

On the atmospheric side of the problem, studies of the costs to specific industries of discharge controls have been rarer. Attention most often has focused on treatment devices (scrubbers, bag houses, electrostatic precipitators) and their applicability to specific source types (boilers, blast furnaces, cement kilns, etc.). Alternatively, work has been done to bring together information on control devices useful for specific residuals, such as dust, nitrous oxides (NO_x), sulfur oxides (SO_x), and so on.[15]

One example of a study based on an industry is that done for the National Air Pollution Control Administration (NAPCA) on wood pulping, by a group of consulting engineers.[16] This study provides very detailed

[13] Ibid., vol. 3, no. 5, "Petroleum Refining."

[14] George O. G. Löf and Allen V. Kneese, *The Economics of Water Utilization in the Beet Sugar Industry* (RFF, 1968); and Paul H. Cootner and George O. G. Löf, *Water Demand for Steam Electric Generation: An Economic Projection Model* (RFF, 1965).

[15] See, for example, Sabert Oglesby, Jr., and others, *A Manual of Electrostatic Precipitator Technology*, pt. 2, "Application Areas" (Birmingham, Ala.: Southern Research Institute, 1970), available as PB 196 381 from NTIS, Springfield, Va.; and U.S. Public Health Service, *Control Techniques for Particulates*, AP-51 (January 1969). Similar volumes on control techniques have been published for CO, hydrocarbons and organic solvents, and NO_x—all from stationary sources—and CO, NO_x, and hydrocarbons from mobile sources.

[16] E. R. Hendrickson and others, "Control of Atmospheric Emissions in the Wood Pulping Industry," for the U.S. Department of the Interior, NAPCA (March 1970), available as PB 190 351 from NTIS.

cost and materials balance information based on surveys of actual pulp mills. But it has several disadvantages aside from the fact that it confines itself to atmospheric emissions. First, the analysts do not look at the effects of changing product quality requirements. Second, though the methods surveyed do include several changes in the production process (as opposed to add-on treatment alternatives), the study includes only changes directly affecting residuals discharge control and not alterations that might be induced by changes in relative input prices, for example.[17]

The Battelle Institute has done a number of studies of the airborne residuals problems of the integrated iron and steel industry, and, while these contain a considerable amount of valuable information, they suffer from the same sorts of defects noted about the other studies. The basic study reports materials balances and costs—including "treatment" costs for unspecified types of control equipment—for the major processing stages of an integrated steel mill (coking, sintering, blast furnaces, steel furnaces, and rolling and finishing).[18] This work was extended in subsequent studies that dealt specifically with residuals generation and discharge.[19] There remains, however, a rather loose connection between process selection and treatment method selection so that it would be difficult, if not impossible, to use the data to predict costs and mill configuration for any given set of circumstances (e.g., scrap and natural gas prices, scrap and ore quality, coal sulfur content, and, of course, limits or charges on residuals discharges).[20] Thus, for example, there is some material comparing the costs of wet and dry collection devices when applied to basic oxygen steel furnaces of different sizes; but one must go considerably beyond the study in order to compare the costs for different input mixes of hot metal and scrap.

[17] A study of the pulp and paper industry that overcomes all these deficiencies is currently being undertaken by Bower, Löf, and Hearon as part of RFF's industry study series. See Blair T. Bower, George O. G. Löf, and W. M. Hearon, "Residuals Generation in the Pulp and Paper Industry," *Natural Resources Journal*, vol. 11, no. 4 (October 1971), pp. 605–23 (RFF Reprint 100).

[18] Battelle Memorial Institute, *Final Report on Technical and Economic Analysis of the Impact of Recent Development in Steelmaking Practices on the Supplying Industries* (Columbus, Ohio, 30 October 1964).

[19] See Thomas M. Barnes and others, *A Cost Analysis of Air Pollution Controls in the Integrated Iron and Steel Industry* (Columbus, Ohio: Battelle Memorial Institute, 15 May 1969), also available as PB 184 576 from NTIS; and J. Varga, *A Systems Analysis Study of the Integrated Iron and Steel Industry* (Columbus, Ohio: Battelle Memorial Institute, 15 May 1969), available as PB 184 577 from NTIS.

[20] One more specific flaw is that sulfur balances are provided for only a few of the process-unit, input-mix combinations discussed in these studies.

The Council on Environmental Quality is now sponsoring a series of industrial impact studies that also address air and water residuals problems.[21] These studies take as given proposed discharge standards, fuel quality specifications, and so on, and include attempts to estimate who will pay the implied costs. Since they involve specific estimates of price changes and of costs borne by the industries themselves, they are clearly a step in the right direction.

The model described in the present study is a more efficient tool for analyzing the costs of various possible residuals management policies because it is capable of the following:

1. Dealing with the three major forms of residuals, with any number of individual residuals within those forms, and reflecting the trade-offs between and among forms implicit in production and treatment choices.
2. Showing the impact on residuals generation and discharge of changes in such underlying conditions as available technology, input costs and qualities, and output quantity and quality requirements.
3. Predicting how the modeled plant would react to various attempts to directly influence its residuals discharges, under the assumption that management acts to maximize profits. (This assumption may appear overly restrictive to many.) In particular, either discharge constraints (allocations) or effluent charges may be applied to any one or a combination of the residuals and the effects observed.

These results in turn may be interpreted and displayed in a number of ways, including:

1. The relation between effluent charges and the amount of discharge reduction, of percentage reduction, or of remaining discharge.
2. The total (or marginal) cost associated with any level of discharge or of discharge reduction for one or more residuals.
3. The relation between the costs associated with any level of discharge or discharge reduction and the refiner's total costs per barrel of crude oil input or, under more tenuous assumptions, the effect of

---

[21] See, for example, Stephen Sobotka and Co., *The Impact of Costs Associated with New Environmental Standards upon the Petroleum Refining Industry* (23 November 1971), pts. 1–3.

such costs on the price of motor gasoline.[22] The latter may be the most meaningful to the public at large, since the public's major point of contact with the refining industry is the gasoline pump.

Models of this type, when constructed for other industries, will allow the exploration of these same questions. The choice of a meaningful basis for reporting and displaying the answers will depend, of course, on the audience to be addressed.

REGIONAL RESIDUALS MANAGEMENT DECISIONS: CONSISTENCY OF AMBIENT AND EMISSION STANDARDS AND BEYOND

As mentioned earlier, present government procedures do not insure even the consistency of emission and ambient standards. In addition there are no methods in use for systematically exploring the cost effects of alternative possible ambient standards or for incorporating existing knowledge of damages from ambient concentrations into policy analysis.[23] Therefore a second significant role for industrial models of the type described here is as a component of a larger regional model. The larger model should also be capable of dealing with airborne, waterborne, and solid residuals simultaneously and, because of the nature of the industrial models, will reflect a range of alternatives for reducing residuals discharges to the regional environment. These alternatives should include changes in input mix, by-product recovery, and the recirculation of resid-

[22] In order to pass from overall refinery cost increases to an increase in the cost of gasoline, all the increases must be allocated to the single, albeit dominant, product. The resulting estimate is, then, an upper limit (ignoring the uncertainty of the total cost figure itself). If some assumptions are then made about the nature of the demand and supply curves in the gasoline market, the average increase in the price of gasoline can be estimated.

[23] This is not to say that such methods have never been used or that the government is not working to develop improved methods today. In setting BOD emission standards for the Delaware estuary area so as to insure meeting chosen ambient standards in a modified least-cost way, a model of the estuary system analogous to the regional model described here was used. Currently EPA is working to make operational the Implementation Planning Program for the study of regional air quality policies. This package is also roughly analogous to the regional model developed at RFF: see Walter O. Spofford, Jr., Clifford S. Russell, and Robert A. Kelly, "Operational Problems in Large-Scale Residuals Management Models," (paper prepared for the Universities–National Bureau Committee Conference on Economics of the Environment, University of Chicago, 10–11 November 1972). See also TRW Systems Group, *Air Quality Implementation Planning Program*, vol. 1, *Operator's Manual*, and vol. 2, *Programmer's Manual*, prepared for EPA (Washington, November 1970), available as PB 198 299 and PB 198 300 from NTIS.

uals-bearing streams in addition to end-of-pipe treatment methods. Models of the regional environment—aquatic, meteorological, and perhaps terrestrial—then should take a particular set of discharges (identified by place, quantity, and residual type in the static model) and transform them into ambient standards and other indicators of the state of the regional environment, such as fish and algal populations. The evaluation of the ambient concentrations, either in relation to standards or through damage functions when these are available, is the final component of the regional model.

Figure 1 is a sketch of the structure of such a regional model and shows how models of industrial residuals dischargers fit in.[24] Two points are of particular importance in understanding it.

First, the residuals generation and discharge part of the model is an optimization problem itself, with costs, prices, and, most important, trial values of effluent charges providing the driving force at any iteration of the overall model. Second, the discharges from the generation and discharge submodel at any iteration are routed through the physical, chemical, and biological models of the regional environment to produce ambient concentrations (e.g., SO_2 in the atmosphere), species populations (e.g., sport fish populations in the regional watercourses), and other pertinent measures such as dustfall per unit area. These environmental measures are then evaluated in the model section containing constraints or damage functions, or both. In the case of constraints the evaluation is on the basis of penalty functions that produce very high pseudo-damages when their constraints are violated.[25] Whether constraints or damage functions are used, the model permits calculation of the marginal damages or penalties assignable to each discharge of each residual. These marginal external costs are then applied as trial effluent charges to their respective discharge activities and the generation and discharge model is solved again. If the overall program is written to include limits on the step size allowable between each iteration, and if the functions are rea-

---

[24] For a detailed description of the framework and computational method, see Clifford S. Russell and Walter O. Spofford, Jr., "A Quantitative Framework for Regional Residuals Management," in Kneese and Bower, eds., *Environmental Quality Analysis*, pp. 115–79.

[25] Note that constraints may be placed on ambient concentrations directly or on some function of one or more of these concentrations. For example, given information on the relation between SO_2 and suspended particulate concentrations and morbidity or mortality, I might want to constrain the latter rather than the former. See, for example, Lester Lave and Eugene Seskin, "Air Pollution and Human Health," *Science*, vol. 109, no. 3947 (21 August 1970), pp. 723–33.

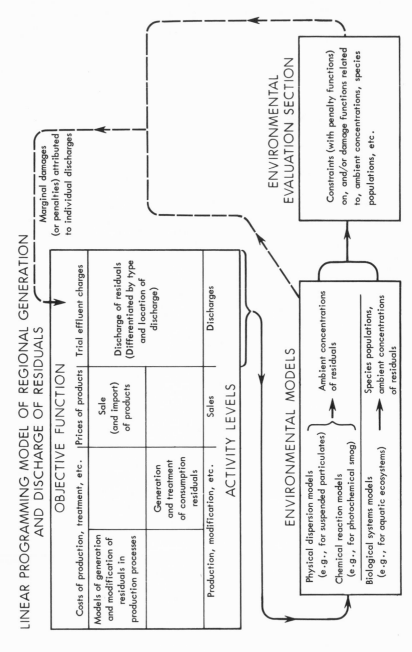

Figure 1. Schematic of the regional residuals management model.

sonably shaped, this process will converge to a local optimum for the given objective function and constraint set (including the environmental quality constraints).[26]

The original rather crude versions of the industrial model type described in this study were constructed in connection with the development of this regional framework. Present efforts to deal with other industries and to construct condensed versions[27] are proceeding as part of an empirical application of this framework to the Delaware estuary region.

The remainder of this book involves a description of the construction and manipulation of a model of an actual industrial operation—a modern petroleum refinery. In chapter 2 I discuss the framework in general terms and some of the techniques available and the difficulties faced. Chapters 3 and 4 involve an examination of the refinery's outputs, petroleum processes, residuals generation and residuals modification alternatives, and methods and sources used in constructing linear programming vectors for these many activities. Here I also begin to show how choices on the petroleum side affect residuals generation and how choices of modification techniques affect discharges. In chapter 5 I establish a benchmark with which to compare later results by examining in detail the results of running the model for a set of base conditions and by showing that these results are in broad agreement with published information on product and process mix and residuals discharges. Chapter 6 is a discussion of observations of the effects of varying such indirect influences on residuals discharge as input availability and price, process technology, composition of output mix, and output quality requirements. Chapters 7 and 8 illustrate the results of varying direct influences on residuals discharges, both in the form of effluent charges and emission constraints. Finally, in chapter 9 I return to the themes of cost estimation and regional modeling in order to draw together the results of construction and manipulation of the refinery model.

---

[26] Whether or not this local optimum is a global optimum depends, of course, on the shape of the model's response surface. The existence of economies of scale or their analogs, as discussed in chap. 2, will generally imply that multi-peaked surfaces can be anticipated. It is then necessary to devise tests for other optima by random starts or other techniques.

[27] See chap. 9.

II

THE MODEL: CONCEPT AND
WORKING STRUCTURE

The fundamental idea behind the conceptual model is that residuals generation may fruitfully be considered as an input to the production process and treated symmetrically with other inputs. Thus the quantity of residuals generated will depend on the relative prices of all the production inputs, including the cost of residuals disposal, that is, the price of "supplying" a certain quantity of residuals generation. The simple production process diagrammed in figure 2 makes this idea more concrete. Heat supplied by burning a purchased fuel, such as natural gas or residual fuel oil, is applied to a liquid stream in a boiler. The heated liquid is fed into a reaction vessel where it comes in contact with a catalyst and undergoes a chemical change. Assuming for simplicity that the reaction is 100 percent complete and that the new liquid is homogeneous, the next step is to cool the output of the reactor vessel to the temperature necessary for storage or transport or whatever. This may be done either by transferring all the heat to cooling water or by transferring some to cooling water and some to the incoming liquid.

In order to make the problem as simple as possible, assume that the temperature at which the reaction takes place, the time the reaction requires, and the amount of liquid to be reacted each day are all given. A firm constructing such a process faces the following two major decisions concerning its input mix:

1. The efficiency of the boiler.
2. The extent to which heat exchange equipment is included in order to transfer heat from the product to the input stream.

14

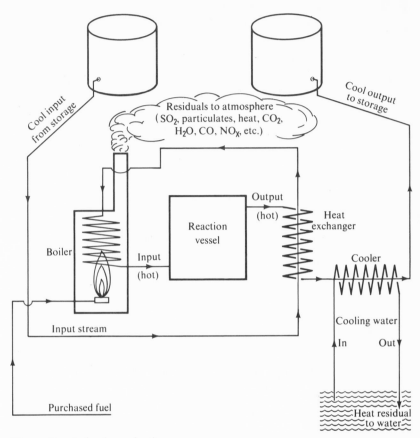

Figure 2. A simple production process.

In deciding on boiler efficiency, the firm chooses how many British thermal units (Btu) it will have to supply in the form of fuel in order to transfer 1 Btu to the input liquid. Hence, given the specific heat of the liquid and the heat content of the fuel, the choice may be translated into one of the quantity of the fuel to be burned per degree rise in liquid temperature. Since burning the fuel involves the generation of the residuals of combustion (CO_2, H_2O, SO_2, NO_x, and particulates), and since the heat not transferred to the input liquid is lost to the atmosphere up the stack or through the furnace walls, the choice of boiler efficiency also involves the choice of residuals generation per Btu transferred. The necessity for choice arises, of course, because more efficient boilers cost more than less efficient ones.

The second decision, on how much heat exchange equipment to buy, again involves money (capital), fuel, and residuals generation. When more capital is employed in the form of heat exchangers, more heat is transferred from the product to the input stream; hence less fuel is required to be burned, at given boiler efficiency, to raise the incoming liquid to the desired reaction temperature[1] and smaller quantities of combustion residuals are generated. In addition, the more heat transferred to the input, the less residual heat there is to be removed by cooling water in the cooling unit; hence the smaller the waterborne energy residual generated.[2]

In the absence of action by public authorities to control discharges of residuals to the environment, the firm would choose a mix of fuel, capital, and residuals-generation inputs so as to minimize the cost of performing the desired operation. In the broader context, in which more options are open, the firm would maximize profit—perhaps by increasing or decreasing the quantity of product to be made by this process or by eliminating the process altogether (by using a different technology to obtain the same product from the same or a different input type or by producing a different product mix entirely).

In the simpler situation, it is clear that:

1. Residuals generation will depend on the relative prices of purchased heat (fuel), boiler efficiency, heat exchange capacity, cooling capacity, and cooling water.
2. In general the solution may include some energy recovery (in the heat exchangers) even in the absence of any public actions to control discharges of combustion residuals or waterborne heat loads.

When considering the broader situation involving possibilities of changes in input type, product type and quality, production technology, and so forth, it is clear that the choice of input mix, including the choice of residuals generation, will depend on a very large number of influences.

Up to this point, it has been assumed that the residuals generated could be disposed of at zero cost. Thus the price of the input, residuals generation, has been zero and the hypothetical firm has had no direct

---

[1] Clearly all these considerations are intertwined and the boiler efficiency choice is not independent of the choice of heat exchanger size. I state it this way to bring out my point without undue complication.

[2] The output stream might be air cooled, but this choice of receiving medium is not important to the fundamental point and is ignored for now.

incentive to economize in its use. The fact that residuals generation might be reduced by the installation of heat exchangers would be due entirely to consideration of the cost of fuel, boilers, and so on. But—to add a further complication—if a public authority puts a charge or a limit on the discharge of one or more residuals, the situation changes. Now the cost of supplying a certain quantity of residuals generation, under a discharge limit, is the cost of reducing the amount generated to the amount the firm is allowed to discharge. In the case of the emission charge, the cost of supplying residuals generation is the least-cost combination of discharge reduction measures and charge payments. Under these new conditions the optimal mix of inputs would have to be determined again, taking account of the cost of residuals generation implied by the public management actions. In the simple model, if residuals discharge limits were put on particulates and SO_2 emissions to the atmosphere and on waterborne heat, the optimal input mix would depend on all the former considerations plus the costs involved in removing SO_2 and particulates from stack gases (*including* any costs of disposing of the residuals generated in these processes themselves—e.g., captured particulates that constitute a solid residual) and the costs of installing cooling towers, spray ponds, or other methods of reducing heat discharges to water.[3]

One can conceive of the process of plant and unit design, then, in either of two ways:

1. There could be two groups of engineers (and accountants) involved and the process would be an iterative one. The group responsible for the design of the process unit itself would consider fuel cost and boiler and heat exchanger cost and, perhaps initially, would assume a zero cost for the supply of residuals generation. The resulting generation would become the problem of the design team responsible for "utility services." This team would then design a least-cost disposal system to meet the existing (and probably the anticipated) limits on residuals discharges facing the plant, given the quantities generated.[4] The marginal costs of such a system that are attributable to the specific residuals would be the appropriate price for the process design team to attach to residuals generation for

[3] Notice that cooling towers that allow the recirculation of cooling water and reduce discharges of heat to water essentially to zero simply transfer that discharge to the atmosphere directly. Waterborne heat discharges are eventually all transferred to the atmosphere, too, but in the process considerable havoc may be wrought with aquatic ecosystems in the neighborhood of the discharge.

[4] Alternatively, of course, they would design a disposal system to minimize the sum of costs and charge payments, if the plant faced a charge schedule.

another iteration of the input-mix determination problem.[5] In any practical situation, such an iterative procedure would probably be carried for only a very few steps, but in principle it could be carried until some arbitrarily small difference in the input mix resulted from another round, assuming the cost functions and other key parts of the problem were reasonably shaped.[6]

2. As the description above suggests, however, an alternative way of viewing the design task is as a programming problem. The functions involved are, of course, inherently nonlinear, and if one wishes to capture them fully the resulting model is complex and difficult to solve even for the simple version discussed above in which size, technology, and desired output are given.[7]

To avoid this problem my approach in constructing working models was to attempt to identify a relatively small set of discrete production alternatives, to cast these in the form of unit activity vectors (that is, vectors giving inputs and cost required for the production of one unit of an input or output of interest), and to condense the two-phase decision process into a single phase represented by a linear program. The objective of the firm may be taken to be profit maximization, cost minimization for given output, or any other convenient variant. The constraint set may include limits on input availability; quality requirements to be met by products; limits on discharges of one or more residuals; and, most importantly, continuity conditions (or mass and energy balance equations) requiring, for example, that the full amount of each residual generated be accounted for explicitly either by intake to a treatment or transport process or by discharge.

To say that once a quantity of residual is generated in the model, it must be treated, transported, or discharged is not to say that *all* residuals must be included in the model. For many purposes, for example, the CO_2 and H_2O residuals from combustion processes will not be of interest and may be ignored in the construction of a response model. Which residuals are of interest will depend to a large extent on the spatial and time dimensions of the study. Thus CO_2 may be a problem on the global, long-term

[5] Problems of *determining* marginal costs for residuals modification units that produce reductions in more than one residual (while, of course, producing more of others) have been ignored.

[6] Shape problems are discussed more fully later in this chapter.

[7] In material prepared for RFF, J. Hayden Boyd and Herbert Mohring described a simple process system of this kind, and the resulting model was solved under various sets of assumptions.

scale but is generally not considered a problem for local areas over the next decade or so.

I use the term "secondary residuals" to refer to residuals generated in such auxiliary processes as treatment, recirculation, and by-product production, as opposed to those generated in the production process itself (the "primary residuals"). This distinction is made for convenience only, for, as described below, secondary residuals are subject to the same continuity conditions as are primary residuals.

The Working Model

The general form of the model is shown schematically in figure 3 and may be interpreted as a familiar linear programming problem:

$$\max c'x$$
$$\text{subject to } Ax \leq b$$
$$\text{and } x \geq 0.$$

The activity levels (x) that are to be chosen by the solution process are shown across the top under six headings: production alternatives $(X_1 \ldots X_H)$; by-product production $(B_1 \ldots B_J)$; materials recovery $(W_1 \ldots W_K)$ from residuals; treatment and transport of residuals $(T_1 \ldots T_L, V_1 \ldots V_M)$; discharge of residuals $(D_1 \ldots D_G)$; and sale of products $(Y_1 \ldots Y_N)$. Corresponding to this division of the possible activities is the vertical division of the A-matrix and the objective function. Thus, for example, the objective function entries corresponding to the production activities are the costs of production; those corresponding to residuals discharges are zero unless some effluent charge is being levied.[8]

The horizontal division of the A-matrix indicates broadly the type and form of the constraints included. Thus each unit of production using process X_1 requires a vector of inputs represented by $(-\tilde{p}_{X_1})$. Total

---

[8] This chapter is a discussion of the features of the model specifically designed for dealing with residuals generation and subsequent handling, but for real industrial applications the section of the model including alternative production processes may be extremely complex in itself. See the diagram and description in chap. 3 of the oil refinery to which these techniques were applied. This part of the problem, however, is also covered in detail in other sources. See, for example, Alan S. Manne, *Scheduling of Petroleum Refinery Operations* (Harvard University Press, 1963).

Rows	Production alternatives $X_1 \cdots X_H$	By-product production $B_1 \cdots B_J$	Raw material recovery $W_1 \cdots W_K$
Production and sale	$+\tilde{e}_{X_1} \cdots +\tilde{e}_{X_H}$	$+\tilde{b}_1 \cdots +\tilde{b}_J$	
Input availability	$-\tilde{p}_{X_1} \cdots -\tilde{p}_{X_H}$		$+\tilde{w}_1 \cdots +\tilde{w}_K$
Output quality	$+\tilde{q}_{X_1} \cdots +\tilde{q}_{X_H}$		
Primary residuals	$-\tilde{r}_{X_1} \cdots -\tilde{r}_{X_H}$	$+\tilde{e}_{B_1} \cdots +\tilde{e}_{B_J}$	$+\tilde{e}_{W_1} \cdots +\tilde{e}_{W_K}$
Secondary residuals		$-\tilde{r}_{B_1} \cdots -\tilde{r}_{B_J}$	$-\tilde{r}_{W_1} \cdots -\tilde{r}_{W_K}$
Possible discharge constraints			
Objective function	Costs of production	Costs of production	Costs of recovery

Figure 3. Schematic of models of industrial residuals management. The $\tilde{e}$ are column vectors of zeroes and ones. A particular vector $\tilde{e}$ has the number of row elements corresponding to the constraint set in which it appears. The occurrence of ones is de-

input requirements for production at level X_1 are $-\tilde{p}_{X_1} \cdot X_1$, and the input availability constraints simply say that

$$\sum_h (-\tilde{p}_{X_h}) \cdot X_h \geq -\tilde{P},$$

or that no input may be used beyond its level of availability as given by the applicable entries in the right-hand side or b-vector. The minus sign on inputs is used here primarily to emphasize the symmetry between traditional inputs and primary residuals generation. In practice, signs may be chosen for computational ease (for example, to avoid negative entries in the b-vector) so long as row consistency is maintained. The other constraints imposed include the following requirements:

1. That all products sold actually be produced.
2. That output quality conform to certain standards.
3. That all residuals generated be accounted for by material recovery, by-product production, treatment, transport, or discharge.

Treatment and transport of residuals $T_1 \cdots T_L, \quad V_1 \cdots V_M$		Sale of products $Y_1 \cdots Y_N$	Discharges of residuals $D_1 \cdots D_g \cdots D_G$	Right-hand side
		$-\tilde{e}_{Y_1} \cdots -\tilde{e}_{Y_N}$		$\geq$ 0
				$\geq$ $-\tilde{P}$
				$\gtrless$ $\tilde{Q}$
$+\tilde{e}_{T_1}$	$+\tilde{e}_{V_1} \cdots$		$\cdots +\tilde{e}_{D_1}$	$=$ 0
$-\tilde{r}_{T_1} \cdots +\tilde{e}_{T_L} \cdots$ $-\tilde{r}_{TL}$	$+\tilde{e}_{V_M}$ $-\tilde{r}_{V_1}$ $-\tilde{r}_{V_M}$ Etc.		$+\tilde{e}_{D_g}$ $\cdot$ $\cdot$ $\cdot$ $+\tilde{e}_{D_G}$	$=$ 0 $\cdot$ $\cdot$ $\cdot$ $=$ 0
		$+1$	$+1$ $+1$	$\leq$ $\tilde{F}$
Costs of treatment and transport		Prices of output	Possible effluent charges	

termined by the function of the column in which the vector appears. Thus in $\tilde{e}_{X_1}$, a one appears in the row corresponding to the output of process X_1. For further discussion of the structure of these constraints, see the text.

The quality constraints for production (requirement 2) at levels $X_1 \ldots X_H$ are shown in the simple form

$$\sum_h \tilde{q}_{X_h} \cdot X_h \gtrless \tilde{Q}.$$

Because such constraints will generally be placed on the concentration of one or another substance in the product, and because final products often result from blending separate intermediate stocks, the actual construction of these rows will generally be more complicated. The method discussed below for including effluent concentration constraints may be applied to such product quality requirements.

Requirement 3, accounting for residuals generated, deserves more detailed discussion. The choice of a by-product production, material recovery, treatment, or transport option for dealing with some residual implies the generation of one or more secondary residuals. These are basically of three types: (1) that portion of a residual subject to treatment that is not removed or altered by the treatment process (as the fly ash continuing up the stack after electrostatic precipitation); (2) the new forms of residuals created by a treatment process (as the sludge from primary and secondary biological treatment of waterborne oxygen-demanding compounds, or the water vapor and CO_2 from refuse burning); and (3) the same residual at a different *place* (as when sewage is piped elsewhere before discharge).[9] The method of inclusion of secondary residuals and of the requirement that they in turn be treated, transported, or discharged is basically straightforward and might be called "row transfer." For example, consider a treatment activity, say l, designed to remove some percentage, b, of residual i from the stream containing it, while in the process generating a quantity of residual, say $r_{T_l j}$, of a secondary residual, j, all at a cost c_{T_l} per unit of i taken into the process (*not*, in this method, the cost per unit of i removed). The activity vector for this process would be:[10]

Row entry description	Treatment activity (T_l) vector entries
Quantity of residual i taken in to treatment process	$+1$
Quantity of residual i *not* removed	$-(1-b)$
Quantity of new residual, j, generated per unit intake	$-r_{T_l j}$
Objective function (cost of unit activity level)	$-c_{T_l}$

Thus if this treatment activity operated at the level T_l is sufficient to account for all the residual i assumed generated, the quantity $(1-b)T_l$ of i would be unaffected, requiring further treatment or discharge; and the quantity $r_{T_l j} \cdot T_l$ of the new residual j would now also have to be accounted for. It is interesting to follow this hypothetical residual, i, through the constraint matrix, using the notation of figure 3. Figure 4

---

[9] In a model in which time enters explicitly, storage of residuals for discharge under more favorable environmental circumstances or in more even quantity would clearly be an important alternative. In that case another type of secondary residual would be defined—one differing in time (*and* in type and place, perhaps).

[10] I concentrate here on tracing residuals flows, but the treatment and transport processes will also require *inputs*, such as steam, water, chemicals, and electricity, which will in turn influence directly the plant's generation of residuals. These inputs must also be included in the activity vectors in order that these influences be felt. Including only the costs of such inputs would not allow reflection of their direct effects on residuals loads.

Activity Levels

	Residuals generation		Treatment		Transport		Discharge				
	X_f .... X_k		T_l .... T_m ...		T_n .... V_o ...		D_i .... $D_{i'}$ .... D_j .... D_h .... D_g				
	X_f	X_k	T_l	T_m	T_n	V_o	D_i	$D_{i'}$	D_j	D_h	D_g
row i	$-r_{X_{f_s}}$	$-r_{X_{ki}}$	$+1$	$+1$			$+1$				
row i′			$-(1-b_l)$	$-(1-b_m)$				$+1$			
row j			$-r_{T_{li}}$	$-r_{T_{mi}}$	$+1$	$+1$			$+1$		
row h					$-r_{T_n^h}$					$+1$	
row g						$-r_{V_{og}}$					$+1$
Objective function	$-c_{X_f}$	$-c_{X_k}$	$-c_{T_l}$	$-c_{T_m}$	$-c_{T_n}$	$-c_{V_o}$	$*$	$*$	$*$	$*$	$*$

$$= \quad \cdots \cdots \cdots \; 0 \; \cdots \cdots \cdots$$

Figure 4. Residuals handling in the linear model. The asterisks indicate that prices (effluent charges) may be applied to the discharge activities.

shows the necessary matrix entries, the activity level designations, the right-hand side, and the objective function.

Treatment activities l and m remove, respectively, the fractions b_l and b_m of the amount of residual i to which they are applied. Thus for every unit of i taken into process l, $(1 - b_l)$ units remain for discharge, assuming there is no opportunity for further treatment. Also, for every unit of i taken in, the treatment processes produce, respectively, $r_{T_{lj}}$ and $r_{T_{mj}}$ units of a new residual, j. This, in turn, must be treated in process n, transported by process o, or discharged. And finally, both processes n and o produce further residuals requiring discharge.[11] The individual row conditions for this example can be written:

$$-X_f \cdot r_{X_{fi}} - X_k \cdot r_{X_{ki}} + T_l + T_m + D_i = 0$$

$$-T_l(1 - b_l) - T_m(1 - b_m) + D_{i'} = 0$$

$$-T_l \cdot r_{T_{lj}} - T_m \cdot r_{T_{mj}} + T_n + V_o + D_j = 0$$

$$-T_n \cdot r_{T_{nh}} + D_h = 0$$

$$-V_o \cdot r_{V_{og}} + D_g = 0$$

In the absence of effluent charges the contribution to the objective function of this section of the problem is given by

$$-(X_f \cdot c_{X_f}) - (X_k \cdot c_{X_k}) - (T_l \cdot c_{T_l})$$
$$- (T_m \cdot c_{T_m}) - (T_n \cdot c_{T_n}) - (V_o \cdot c_{V_o}).$$

One extension of this general method is worth specific mention, for it allows one to deal with the common situation in which several streams (for example, process water streams from different processing units) contain a number of residuals in different proportions and are subject to several possible treatment stages, each removing a particular proportion of each residual. The basis of this method is to use, in place of any single residual quantity, the quantity of the carrying stream as the variable in the required row continuity conditions. Only at the point of discharge are the concentrations of residuals used to obtain quantities of residuals explicitly. Thus, assume there is interest in the residuals discharge problems involving two water streams of volumes V_1 and V_2, generated in

---

[11] Usually a transport activity simply changes a residual's location and not its form or amount. Thus the entry $-r_{V_{og}}$ might just as well be equal to -1. This may strike the reader as trivial in a model of a single plant, but the distinction between discharge locations is far from trivial in the broader context of regional residuals management decisions.

production processes X_1 and X_2 respectively. Assume that both contain three residuals of interest, in concentrations O_{11}, O_{12}, and O_{13} and O_{21}, O_{22}, and O_{23} respectively. Further assume that each stream may be discharged directly; subject to treatment process A and then discharged; or subject to process A followed by process B and then discharged. Process A is assumed to remove fractions a_1, a_2 and a_3 of the three residuals; process B removes fractions b_1, b_2 and b_3 of the remaining quantities of the residuals.[12] For simplicity, all process inputs and any secondary residuals generated in A and B are neglected. Then the required matrix entries may be written as in figure 5. The total discharged quantities of residuals 1, 2, and 3 are simply obtained as the sum of discharges 1 through 6.[13]

The most obvious problem with this approach involves economies of scale. If, in the program, any subset of the several streams may be chosen for treatment and others discharged, the total volume for which the facility must be designed can never be known in advance. Hence unit costs applied to A and B in the objective function must necessarily be arbitrary. One can guess at the probable volume, choose the largest possible volume, or adopt some other strategy, but whatever is done will have some effect in turn on the volume actually determined in the solution. As long as there are economies of scale, these guesses will tend to be self-fulfilling, because large assumed volume will imply low unit costs that will in turn encourage the wide adoption of treatment (and vice versa). In the model described in this study, unit treatment costs are based on the size of the treatment plant that would be required to handle the entire volume of process water for the benchmark refinery. More subtle problems may arise because of nonconvexities introduced by variations in the removal per unit cost between A and B for one or more residuals. These are discussed below.

Having explored in some detail how the linear response model reflects residuals handling alternatives, I now go on to describe how it can be used to investigate industrial response to residuals management action

[12] The assumption that removal "efficiencies" are independent of relative and absolute concentrations may be inaccurate in certain cases (as BOD and phenols or BOD and turbidity). If the inaccuracy is unacceptable, it may be possible by exercising considerable ingenuity to salvage the linear programming framework through the use of piecewise approximations to nonlinear functions.

[13] This may be done by introducing three new activities D_7, D_8, and D_9 with $+1$ entries respectively in the rows for residual 1, residual 2, and residual 3 and by requiring that the sum across each of those rows be zero. D_7 will then be the total discharge of residual 1, and so forth.

Activity levels

Rows	A_1	A_2	B_1	B_2	D_1	D_2	D_3	D_4	D_5	D_6
Vol. 1 (from production)	+1				+1					
Vol. 1 (from process A)	−1	+1				+1				
Vol. 1 (from process B)			−1				+1			
Vol. 2 (from production)		+1						+1		
Vol. 2 (from process A)		−1	+1						+1	
Vol. 2 (from process B)				−1						+1
Residual 1 discharged					$-O_{11}$	$-(1-a_1)O_{11}$	$-(1-b_1)(1-a_1)O_{11}$	$-O_{21}$	$-(1-a_1)O_{21}$	$-(1-b_1)(1-a_1)O_{21}$
Residual 2 discharged					$-O_{12}$	$-(1-a_2)O_{12}$	$-(1-b_2)(1-a_2)O_{12}$	$-O_{22}$	$-(1-a_2)O_{22}$	$-(1-b_2)(1-a_2)O_{22}$
Residual 3 discharged					$-O_{13}$	$-(1-a_3)O_{13}$	$-(1-b_3)(1-a_3)O_{13}$	$-O_{23}$	$-(1-a_3)O_{23}$	$-(1-b_3)(1-a_3)O_{23}$
Objective function	$-c_A$	$-c_A$	$-c_B$	$-c_B$						

Figure 5. Method of handling multi-residual streams subject to a variety of treatment processes. A_1 indicates the level of process A applied to stream 1.

on the part of public authorities involving effluent charges, discharge quantity constraints, or even discharge concentration limits. The effects of these direct measures on discharges, costs, production volume, and so on are determined. At another level, indirect influences such as process technology, requirements for output quality, available input quality, and relative input prices may be varied, and under each set of these influences the impact of direct residuals management actions may be examined. Thus if there is good reason to expect changes in any of the indirect influences, this model permits investigation of the residuals generation and discharge pattern of the firm under the new conditions as well as under present conditions.

In figure 4 the place of effluent charges in the model has already been indicated. Since discharges are explicit activities and not simply slack variables, effluent charges fit perfectly as the unit activity costs. If effluent charges, specific to residuals and to locations, were applied to the discharges in figure 4, the new objective function would include

$$-(D_i + D_{i'})\alpha_i - D_j\alpha_j - D_h\alpha_h - D_g\alpha_g{}^{14}$$

where α_i is the fee per unit of discharge of residual i. If the charges were not specific to location, but only to type, the result would be

$$-(D_i + D_{i'})\alpha_i - (D_j + D_g)\alpha_j - D_h\alpha_h$$

since by hypothesis residual g differed from residual j only in discharge location.

Constraints on discharge quantities are easily included by attaching additional rows. To constrain a specific discharge D_i, to be less than $\bar{D}_i$, a new row is introduced in which the activity (column) D_i has a $+1$ entry. Then a constraint is put on the new row, requiring $D_i \cdot 1 \leq \bar{D}_i$. If several separate discharge activities all involve the same residual type, it is a simple matter to constrain their sum. The task of constraining discharge concentrations (as milligrams of BOD per litre of water) is somewhat more complicated. Referring back to figure 5, consider the possibility of constraining the concentration of residual 1 to be less than $\bar{R}_1$. The total discharge of residual 1 is

---

[14] Throughout this study I ignore the dimension of time, but in a practical application the time pattern of discharges may be very important because of variation in the assimilative capacity of the natural world. Then options to store residuals for later discharge and on-off options for treatment operation become important, and effluent charges may be differentiated by timing as well as by type and location. These dynamic considerations add considerably to the complexity of the model and require different solution techniques.

$$Q = O_{11}D_1 + (1 - a_1)O_{11}D_2 + (1 - b_1)(1 - a_1)O_{11}D_3 + O_{21}D_4$$
$$+ (1 - a_1)O_{21}D_5 + (1 - b_1)(1 - a_1)O_{21}D_6,$$

while the total volume of the discharge stream is $D_1 + \ldots + D_6$. Thus the concentration is

$$\frac{Q}{D_1 + \ldots + D_6}.$$

Assume that it is desired to constrain this concentration to be less than R_1. Thus

$$\frac{Q}{D_1 + \ldots + D_6} \leq R_1.$$

In this form the constraint cannot be included in the linear program, but by clearing fractions and subtracting the right from the left side, a more useful form is obtained:

$$D_1(O_{11} - R_1) + D_2[(1 - a_1)O_{11} - R_1] + \ldots$$
$$+ D_6[(1 - b_1)(1 - a_1)O_{21} - R_1] \leq 0.$$

Thus the row entries become the differences between the actual and the desired concentrations, and the constraint simply says that the volume-weighted average of these differences must be less than or equal to zero.[15] The difficulty with this technique is that exploration of alternative levels of the concentration constraint, R_1, is cumbersome and time consuming, since in general the only way to proceed is to insert an entire new set of row entries in the vectors $D_1, \ldots, D_6$. (In the refinery, when dealing with a large number of alternative gasoline-blending stocks or a large number of process-water streams, this becomes a significant problem.) Only if it is possible and meaningful to require that the total volume remain the same for all solutions can the level R_1 be varied easily. To see this, define

$$Z \equiv D_1(O_{11} - R_1) + D_2[(1 - a_1)O_{11} - R_1] + \ldots$$
$$+ D_6[(1 - b_1)(1 - a_1)O_{21} - R_1]$$

so that the constraint requiring at most concentration R_1 becomes simply $Z \leq 0$. Further, define

$$\bar{D} \equiv D_1 + \ldots + D_6$$

[15] This method has been adapted from a Harvard Business School discussion problem: "The Tascosa Refinery," ICH 9C47R1, TOA, 4R2, Harvard University, 1965. There it was used to set lead concentration limits on gasoline blended from several stocks.

and require the total volume to be a constant. ($D_1, \ldots, D_6$ can vary individually.) Now consider a new, tougher concentration constraint $R_2 = R_1 - \Delta$. In general the constraint for R_2 would be

$$D_1(O_{11} - R_2) + D_2[(1 - a_1)O_{11} - R_2] + \ldots$$
$$+ D_6[(1 - b_1)(1 - a_1)O_{21} - R_2] \leq 0.$$

But this is equivalent to

$$D_1(O_{11} - R_1 + \Delta) + \ldots + D_6[(1 - b_1)(1 - a_1)O_{21} - R_1 + \Delta] \leq 0$$

or

$$D_1(O_{11} - R_1) + \ldots + D_6[(1 - b_1)(1 - a_1)O_{21} - R_1] + \bar{D}\Delta \leq 0$$

or, finally,

$$Z \leq -\bar{D}\Delta.$$

Thus if constant total volume is guaranteed, a concentration constraint can be varied simply by varying the right-hand side.

A variety of indirect influences on residuals generation and discharge may also be studied by manipulating the values of the objective function, the right-hand side, or the matrix of coefficients itself. Thus in the objective function any of the price or cost figures may, in principle, be altered and the effect observed, though in practice there may be interest only in the price of a key input (such as coal to a thermal electric-generating plant), a particularly important product (such as motor gasoline from an oil refinery), or an actual or potential by-product (such as sulfur in the refinery). On the right-hand side, input availabilities and output quantity requirements may be changed.[16] And finally, advances in production or residuals-handling technology may be reflected by changing coefficients within the A-matrix itself. Such changes may take the form of introducing entire new columns to represent possible new processes. Another alternative is to change one or two coefficients within existing columns to reflect progress in some aspect of a largely unchanged process.

SOME DIFFICULTIES

In the description above several potential sources of difficulty in the techniques for constructing linear models of industrial production, resid-

[16] And quality requirements, subject to the remarks above.

uals generation, treatment and discharge have already been mentioned. One of these was the matter of scale economies, which arises any time unit capital or operating costs vary inversely with the scale of the facility. On the practical level there is no single correct unit price to attach to the activity vector for such a facility,[17] and the usual trick of approximating the nonlinear curve with piecewise linear segments will not work because the segments would not fill up in the correct order in the solution process. But on a more fundamental level, economies of scale make any problem involving a choice of capacity a difficult one because the response surface, corresponding to the falling marginal and average costs, will have multiple optima.[18]

Similar effects are created whenever the marginal cost of obtaining some desired result falls as the amount obtained rises. Thus in standard sewage treatment for removal of oxygen-demanding organics, the cost of BOD removal, when graphed against percentage removal achieved, follows an S-shaped curve, with falling marginal costs of additional removal over a significant range.[19] The determination of the appropriate treatment level is then subject to the same difficulties as is that of the proper size of a facility exhibiting economies of scale. It is possible to set up the constraint matrix entries for a standard treatment plant in such a way that the segments must be chosen in the proper order,[20] but the problem will still involve nonconvexity.[21] In the model reported in this

[17] That is, there is no single correct price until after the problem has been solved and the scale of the facility is known.

[18] In a general programming framework it is at least conceptually possible to deal with the multi-peaked surface by making random starts within the feasible space. See Peter Rogers, "Random Methods for Non-convex Programming" (doctoral dissertation, Division of Engineering and Applied Physics, Harvard University, 1966).

[19] See, for example, Richard Frankel, "Economic Evaluation of Water Quality: An Engineering-Economic Model for Water Quality Management" (University of California at Berkeley, College of Engineering and School of Public Health, SERL Report 65-3, January 1965).

[20] See, for example, Clifford S. Russell and Walter O. Spofford, Jr., "A Quantitative Framework for Residuals Management Decisions," in Allen V. Kneese and Blair T. Bower, eds., *Environmental Quality Analysis: Theory and Method in the Social Sciences* (Johns Hopkins Press for RFF, 1972); and for a different approach, Daniel P. Loucks, *Stochastic Methods for Analyzing River Basin Systems*, Technical Report No. 16 (Cornell University Water Resources and Marine Sciences Center, August 1969).

[21] As mentioned above, where more than one residual is involved in a *single* process, removal efficiencies may be interdependent (for example, BOD and phenols removal in a standard biological treatment plant). This interdependence essentially introduces cross-product terms into the problem and makes the linear model of doubtful value. If, however, interdependencies exist only *between* processes, the problem is potentially amenable to solution. Thus the removal efficiency of electrostatic precipitation of particulates in stack gases is directly related to the SO_2 content of those gases. Hence

book, no evidence was found for multiple optima in the course of solving the model using different starting points.

A related problem arises in modeling the situation in which the decision must be made whether or not to install some residuals treatment equipment and, if it is installed, whether or not to operate it. One may again think of declining marginal cost: first, a very steep cost segment reflects the installation but not the operation of the equipment; then a flatter segment reflects operation costs (once installed) up to capacity. Thus

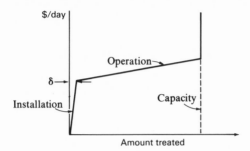

where $\delta \rightarrow 0$.

This problem may be approached through integer or dynamic-programming solution methods, although there is a practical limit in both methods to the number of such alternatives that may be considered. In this study installation was not considered separately from operation. It was assumed that the refinery did not yet exist (a "grass roots" design) and that any restrictions or fees relating to discharges were taken to be constant by the refiner both within and between years over the life of the facility. Hence no equipment would be installed unless it were also to be operated.

With the conceptual framework thus established, it is now appropriate to go on to discuss in some detail the process and residuals-handling alternatives to be included in the refinery model. The two major purposes of the next two chapters are (1) to prepare the reader to understand the results reported in chapters 5 through 8 by familiarizing him with the setup of the refinery, and (2) to give him some idea of how the activity vectors were constructed, what sources were used, what assumptions were made, and so forth.

---

burning low-sulfur coal in a thermal electric plant will reduce SO_2 emissions but may increase particulate emissions for the same equipment and operating policy. This sort of interdependence can be handled within the linear framework through careful definition and linkage of relevant alternatives.

III

DESCRIPTION OF THE PETROLEUM REFINERY MODEL: HYDROCARBON STREAMS AND PROCESS UNITS

Originally there were three reasons for choosing the petroleum refining industry as the one for which to construct an example of the type of model described in chapter 2. First, it is a large industry: refineries are located in many areas of the United States and frequently are major contributors to local environmental quality problems. Second, refineries are sufficiently complex to illustrate many of the interactions discussed earlier. For example, it is possible to show how input price and quality constraints and product quantity and quality constraints affect residuals generation. In addition the model shows how effluent charges and discharge standards affect residuals generation and discharge and how reductions in discharges to one medium may give rise to discharges to another. A third reason for the choice of petroleum refining is the important, practical consideration that there is in the public domain a relatively large amount of technical and cost information concerning refinery processes.

To understand the model described below, the reader needs some minimum knowledge of the nature of the inputs and outputs of a refinery—that is, some knowledge of hydrogen-carbon (or hydrocarbon) compounds and the operations performed on them in the course of refining. Hydrocarbon molecules vary greatly in size, weight, and physical and chemical properties, depending on the number of carbon atoms present (indicated by C_n), the arrangement of these atoms, the nature of the bonds between them, and the resulting number of hydrogen atoms. In general the larger the molecule is (the more carbon atoms present), the

32

more viscous the liquid is, the higher its boiling point,[1] and the higher the temperature to which it must be raised to burn without the continued addition of outside heat (flash point). Crude oils (that is, oil as found in nature) are mixtures of hydrocarbons of widely differing weights and molecular arrangement, with greater or smaller amounts of impurities, such as sulfur, metals, brine, and nitrogen. These differences occur not only between distinct fields, but even within a single field. Relative crude oil prices reflect compositions as they in turn affect final product availabilities and difficulty of refining.

Even without their impurities, however, most crudes are of relatively little use to our present society. Because of their high flash points and viscosities they could generally only be burned in stationary boilers to generate steam for turbines or heating. One of the aims of refining, then, is to separate crude oil into a number of fractions, according to boiling point (and hence other correlated physical characteristics). This separation is performed by boiling the crude oil and condensing the vapors differentially according to boiling point. Because of the enormous importance of the internal combustion automobile engine, the most valuable of these fractions is one consisting of hydrocarbons boiling between 80°F and 450°F, roughly C_5–C_{10}, which are suitable for automobile gasoline. The natural availability of this fraction, while varying from crude to crude, is quite limited. Thus some of the most important processes in modern refining are those designed to reduce the size of long hydrocarbon molecules to gasoline size by splitting or cracking them.[2] This allows the refiner to increase the proportion of product in the motor gasoline range over that available from the crude oil.

Finally, of course, some of the impurities in the original crude must be removed either before or after other processing steps. Removal will often be accomplished first if the impurity tends to increase refining costs (as when sulfur and metals clog catalyst particles and reduce their effectiveness). They may be removed only at the end if their major effect is on the product's acceptability to consumers or on some final process stage. As

---

[1] The very lightest hydrocarbons, those containing one to four carbon atoms (or C_1–C_4 for short), are gases at normal room temperatures.

[2] Note that splitting a long hydrocarbon chain in half is not the same as splitting a single atom (nuclear fission) since individual atoms of carbon and hydrogen are not disturbed and retain their original numbers of protons and neutrons. Only chemical bonds are rearranged.

Polymerization is a process used to increase the size of very short hydrocarbons by combination. It is quantitatively much less important than the cracking-type processes.

one would expect, when crude oil impurities are removed, they frequently become process residuals. The model concentrates on sulfur as an example of a residual originating in the crude and appearing at various points in the refinery as an airborne or waterborne residual.

A petroleum refinery, then, consists of a number of interconnected process units each designed to perform one or more of the three basic kinds of refining operations on liquid and gaseous hydrocarbons: separation by boiling point (fractioning); transformation of the molecules by splitting, combining, or rearranging (e.g., catalytic cracking, reforming, alkylation, and isomerization); or quality enhancement by the removal of impurities (e.g., caustic scrubbing and hydrogen treating).[3] Where molecular transformations are induced, the process unit must also generally have an attached separation unit since the transformation will never be a 100 percent change of homogeneous input stream to homogeneous output stream but will leave some input unaffected and at the same time create more than one new hydrocarbon product. (For examples, see the descriptions below of reforming, catalytic cracking, and coking.) The model includes an array of processes designed to increase the quality and yield of motor gasoline per barrel of crude oil input. It does not include any petrochemical processes or lube oil refining. Some processing of non-gasoline products is included, however, in order to reflect a realistic array of choices of processing combinations.

In this chapter, the flow of hydrocarbons streams is traced through the refinery from crude oil to salable products. Where necessary, the implications are indicated for residuals generation of the process choices available in the model. But not until the next chapter is the residuals-handling system as a whole described and the treatment and recirculation alternatives included.

A few additional assumptions and ground rules may be noted at this point. First, the analysis is based on the assumption that the refinery has not yet been constructed, but that certain decisions have been made that will constrain its size and possible configurations. This approach represents a compromise that allows scope in the inclusion of alternative processes but does not require consideration of the question of optimal size in relation to market demands and scale effects on costs. In particular, it is assumed that there is a limit on the daily availability of crude oil at the

---

[3] Some processes do not fit into this categorization. (For example, separation of desired streams may be accomplished by selective solvents when boiling points are close together.) But almost all major refinery units have one of these three purposes.

refinery. This availability effectively limits the size of the installation as a whole, though it allows some range of sizes for most of the individual processing units. The overall limit on deliveries is 150,000 barrels (bbl) per day, roughly the size of major new refineries recently constructed in the United States.[4]

The hypothetical refinery is allowed to make use of either of two crudes. One is a relatively low-sulfur oil with the characteristics of an East Texas crude; the other is higher in sulfur and represents an Arabian mix oil from the Persian Gulf.[5] In the basic models the higher-sulfur crude is assumed to be significantly cheaper but subject to limited availability. This makes it easy to explore later the impact of the average sulfur content of crude oil on refinery water use and residuals discharge by varying this limit on availability. For the purposes of this study, however, it is not necessary that this limit coincide with "reasonable" limits under the import quota system. But the sensitivity of certain of the results to changes in the relative crude price must be explored.

BASIC PROCESS UNITS

The refinery's processing units and the petroleum streams linking them are shown in figure 6 under the assumption that all possible units are included. Also indicated are points at which choices between processing routes may be made in the model.[6] Because of the complexity of this

[4] Since the model operates with the day as its unit of time, the figures may be thought of as per stream day for every unit. Average throughputs in barrels per calendar day could be obtained by observing the operation of this refinery over a year and allowing for unit downtime.

For the entire United States the average refinery size is 52,510 bbl/stream day. But for the important eastern refinery states of Delaware, Pennsylvania, and New Jersey the average size is 72,660 bbl/stream day, and for Texas it is 84,251 bbl/stream day. These averages include refineries of all types; averages for integrated refineries only would be considerably larger. See *Oil and Gas Journal*, vol. 69, no. 12 (22 March 1971), p. 94.

[5] The crude characteristics, especially fractions and sulfur percentages, have been drawn with some adjustments from M. M. Stephens and Oscar F. Spencer, *Petroleum Refining Processes* (Pennsylvania State University, 1956).

[6] The most important sources used in constructing this model were W. L. Nelson, *Petroleum Refinery Engineering* (McGraw-Hill, 1949); idem, *Guide to Refinery Operating Costs* (Tulsa, Okla.: Petroleum Publications, 1970); Stephens and Spencer, *Petroleum Refining Processes;* S. E. Johnson, M. C. Forbes, and P. A. Witt, "Waste Disposal Cost Allocation" (paper presented at the Rocky Mountain Regional Meeting, National Petroleum Refiners Association, Billings, Mont., October 1968); Bonner and Moore Associates, Inc., *U.S. Motor Gasoline Economics*, vol. 1, prepared for the

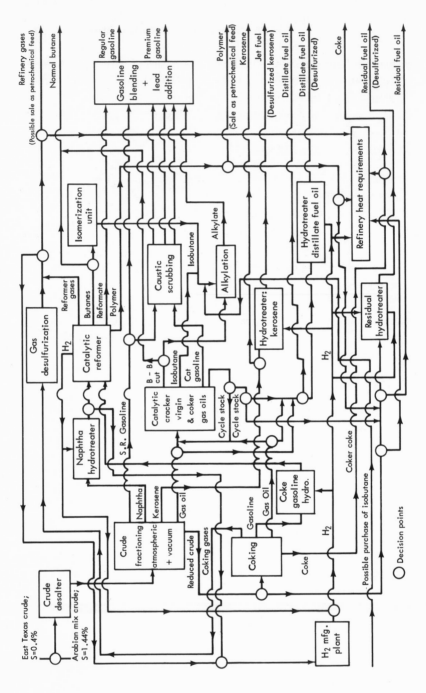

Figure 6. Sample refinery for investigation of residuals generation and discharge: schematic of product flows.

diagram, other figures include simplified schematics of several of the refinery's subsystems: the gasoline production units (figure 7), the hydrogen-hydrotreating system (figure 8), and the refinery gas system (figure 9). The discussion below is keyed to these simplified diagrams.

The Gasoline Production Subsystem

This is by far the most important subsystem, whether one is interested in the quantity of material processes, the value of products, or the generation of residuals. It is not, of course, completely separable from the other systems, especially because the crude desalter and the distillation units are the starting points for all the systems. This discussion, however, passes briefly over these connections, simply referring the reader to the appropriate subsystem in order to stress the several routes from crude oil to finished gasoline.

Desalter.   The first choice, subject to the availability limitation, is the type of crude to be charged: the East Texas low sulfur or the cheaper and higher-sulfur Arabian mix. All crude is first routed through the desalter, which removes saltwater brines found in the oil. Modern practice is to effect this removal by applying electrostatic force, but each barrel of crude is assumed to require 2.4 gallons of makeup water for carrying away the brines, and the total desalter effluent is assumed to be 2.5 gallons per barrel.[7] This stream is assumed to pick up not only brine, but also oil and other residuals through its contact with the crude.

Crude fractioning.   Crude oil from the desalter passes on to the fractioning complex where it is subject to atmospheric distillation, the heavy residue from this step being sent to a second tower for distillation in a vacuum. Vacuum distillation allows the separation of products with relatively high boiling points at lower temperatures than would be required at atmospheric pressure. This reduces the heat input required and prevents unwanted cracking and reforming reactions from taking place during fractioning. Both atmospheric and vacuum distillation are included

---

API (New York, 1 June 1967); M. R. Beychok, *Aqueous Wastes from Petroleum and Petrochemical Plants* (John Wiley, 1967); A. S. Manne, *Scheduling of Petroleum Refinery Operations* (Harvard University Press, 1963); API, *Manual on Disposal of Refinery Wastes: Volume on Liquid Wastes* (New York, 1969); U.S. Public Health Service, *Atmospheric Emissions from Petroleum Refineries: A Guide for Measurement and Control*, PHS Pub. 763 (1960); and the Bechtel Corporation's technical review of an earlier draft of this study, 12 April 1972.

[7] Based on Bechtel technical review.

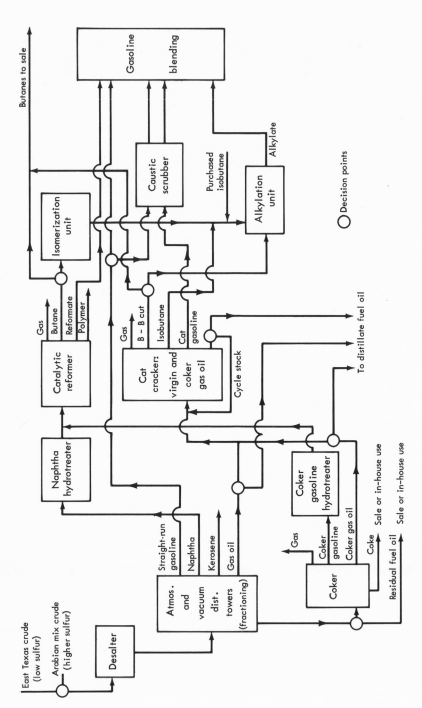

Figure 7. The gasoline production system.

in a single activity in the model, and the products from this two-stage process are shown in table 1 along with the utility requirements, residuals generation, and costs per barrel of crude charged.[8] In order of increasing boiling points and specific gravities, these products are straight-run gasoline, naphtha, kerosene, gas oil (light and heavy), and reduced crude.

As an example of substitution between residuals generation and other inputs, a choice of fractioning processes has been included for each crude. One process uses relatively more capital in the form of heat exchangers for transferring heat from hot product to cool crude oil streams and in the form of more efficient boilers for supplying fresh heat to the crude. In this process, fresh heat input (fuel oil burned in a boiler furnace) and waste heat residual (in the form of heated cooling water) are both relatively low.[9] The second process uses relatively less capital and hence requires more fresh heat input and generates a greater heat residual. Through this choice (and similar ones provided for the coking and catalytic reforming processes) for the refinery as a whole, changes in the cost of fresh heat input have a small but noticeable effect on residual heat generation and total fresh heat input. This will be observed in chapter 6.

In addition to fresh and waste heat, the fractioning unit requires other utility inputs and generates other residuals. As physical flows (as entries

[8] The concentrations of specific residuals in the process condensate are discussed in chap. 3. The process costs are generally based on information provided by Nelson in *Guide to Operating Costs* and include the costs of capital, labor, electric power, and some materials such as catalysts. Requirements for fresh heat inputs, cooling water (residual heat), and stream are not reflected in costs but rather are included explicitly as physical flows to bring out their implications for residuals generation. Capital costs are calculated using an 8 percent interest rate and 20-year life. The chosen interest rate is probably low as a measure of the opportunity cost of capital to the refiner and hence will tend to bias the results toward capital-intensive processes and residuals modification units.

[9] The calculations for these alternatives are based on information on refinery process heat use prepared by J. Hayden Boyd and Herbert Mohring for RFF. The assumptions were that for the low-capital alternative, furnace efficiency was 80 percent and heat exchange accounted for about 40 percent of net heat input to feed stock. For the high-capital alternative, furnace efficiency was assumed to be 85 percent and investment in heat exchangers was assumed to have been increased so that 44 percent of net heat addition was achieved through exchange. Capital costs for heat-related equipment were assumed to be about 30 percent higher for the second alternative. Following Nelson, for each process a significant fraction of the total heat added to the oil is assumed dissipated to the atmosphere. (See, for example, Nelson, *Guide to Operating Costs*, p. 32, on fractional distillation, and p. 71, on platforming where half the heat added is assumed to be dissipated.) Here it is assumed that two-thirds of the heat in the process steam is lost to the atmosphere.

For a discussion of the various possible sources of fresh heat, including purchased fuel and internally generated petroleum streams, see chap. 4.

TABLE 1.   Fractional Distillation: Petroleum Products, Utilities, Residuals, and Costs

Products	East Texas crude[a]	Arabian mix crude[b]
Straight-run gasoline		
Weight (%)	22.3	16.4
Volume (%)	26.0	19.4
Weight per barrel (*lb*)	251	248
Sulfur weight (%)	0	0.04
Research octane number	70	58
Boiling range (°*F*)	100–300	100–300
Naphtha		
Weight (%)	9.4	10.9
Volume (%)	10.0	12.0
Weight per barrel (*lb*)	273	268
Sulfur weight (%)	0.1	0.4
Boiling range (°*F*)	300–400	300–400
Raw kerosene		
Weight (%)	9.9	11.5
Volume (%)	10.0	12.0
Weight per barrel (*lb*)	288	284
Sulfur weight (%)	0.15	0.7
Boiling range (°*F*)	400–500	400–500
Light gas oil		
Weight (%)	20.9	22.6
Volume (%)	20.0	21.0
Weight per barrel (*lb*)	306	303
Sulfur weight (%)	0.25	1.13
Boiling range (°*F*)	500–700	500–700
Heavy gas oil		
Weight (%)	20.3	20.5
Volume (%)	19.0	19.0
Weight per barrel (*lb*)	312	318
Sulfur weight (%)	0.48	2.2
Boiling range (°*F*)	700–950	700–950
Reduced crude		
Weight (%)	17.3	18.2
Volume (%)	15.0	15.5
Weight per barrel (*lb*)	338	345
Sulfur weight (%)	1.28	3.33
Boiling range (°*F*)	>950	>950

Utilities & residuals per barrel	High Capital	Low Capital
Fresh heat input (10^6 *Btu*)	0.1107	0.1274
Steam (*lb*)	45	45
Residual (waste) heat (10^6 *Btu*)	0.0791[c]	0.0845
Condensate (*gal*)	5.4	5.4
Cost per barrel (*$*)	0.080	0.076

Sources: M. M. Stephens and Oscar F. Spencer, *Petroleum Refining Processes* (Pennsylvania State University, 1956), pp. 56–59; and the Bechtel Corporation's technical review of an earlier draft of this study, 12 April 1972.

[a] Weight per barrel (1 bbl = 42 gal) = 292 lb.   Sulfur weight = 0.4%.

[b] Weight per barrel (1 bbl = 42 gal) = 295 lb.   Sulfur weight = 1.44%.

[c] Residual heat *can* be greater than fresh heat because of the heat in the steam. The fresh heat figure given is *gross*. That is, it represents fuel input to the furnace; 15%–20% will be lost to boiler efficiency.

in the process vectors), the model includes process steam required and process waste water volume generated.[10] The steam is generally used in all fractioning steps to improve the refiner's ability to separate the products in the desired way.[11] It may also be used to remove impurities from petroleum or waste water streams and to introduce heat directly into petroleum stocks. Interest here is in the steam that comes in direct contact with petroleum stocks or with other residuals streams (not with whatever quantity is used for tank-heating coils, pumps, etc., which is generally assumed to be clean enough for direct recycling or discharge without treatment).

The steam that comes in contact with product or residuals streams entrains substances resulting in a condensate that is relatively "dirty." These condensates may be referred to as "sour," "foul," or "phenolic," depending largely on the relative preponderance of dissolved hydrogen sulfide (H_2S)—or other sulfur compounds—and phenols.[12] They also are likely to contain some oil and to exhibit some oxygen demand when exposed to the action of bacteria or chemical oxidants.[13] These important residuals-bearing streams are discussed in the next chapter. For now it is enough to note that nearly every process unit will generate a stream of condensate more or less rich in residuals.[14]

Straight-run gasoline. Now, to follow the individual fractions from the crude-fractioning towers—the straight-run gasoline may go directly to

[10] Electrical power is also required for almost all processes for pumping, compressing, and so on. I have included the cost of this power in process costs, but this is not strictly correct if one wishes to trace out all the implications for residuals discharges of various actions applied to the refinery.

[11] See Nelson, *Petroleum Refinery Engineering*, pp. 209–11, for more information on this point. In general, steam requirements for all the refinery processes were taken from idem, *Guide to Operating Costs*, which gives detailed cost and utility data for all major process units.

[12] "Phenol" refers both to a specific compound and to a class of similar compounds. Thus phenol has the formula C_6H_5OH and is a hydrocarbon compound with an attached OH radical. The phenols are similar compounds with different numbers of carbon and hydrogen atoms: the cresols (C_7H_7OH) and the xylenols (C_8H_9OH). Within each of these latter groups a number of different atomic arrangements are possible, hence the plural form of the names. Phenols may be toxic to aquatic life and impart an undesirable taste and odor to water even when present in very low concentrations.

[13] BOD is used to designate this oxygen demand, but chemical oxygen demand is probably the more appropriate measure. For a discussion of the distinction in nature and in laboratory tests, see Beychok, *Aqueous Wastes*, pp. 5–13.

[14] It is assumed for this unit that condensate effluent is equal to process steam introduced. This is generally the approach, but the residuals loadings in the condensates from three of the units are assumed to be negligible, and hence the condensates are not followed through to discharge. This produces for the refinery in the aggregate the same effect as would an assumption of a loss to the atmosphere of 17.4 percent of the steam introduced.

the gasoline-blending section (discussed more fully below) or may first be "scrubbed" with a caustic soda solution to remove sulfur, sulfur compounds like mercaptans, and phenols to improve odor and increase the gasoline's susceptibility to tetraethyl lead (TEL).[15] Octane rating measures a desirable characteristic of motor gasoline, the slowness of its combustion. For high-compression engines, high octanes are necessary to prevent premature combustion and "knocking." The "clear" octane rating is the rating of the blending stock without an addition of TEL.[16] "Susceptibility" to TEL refers simply to the size of the increase in octane ratings with given additions of TEL. Whether or not a gasoline-blending stock must be scrubbed will depend on its source (and hence its quality), on the qualities of the other available blending stocks, and on the quality requirements fixed for the final motor gasoline blends. Whenever caustic scrubbing is used, the major process residual is a stream of spent caustic with varying quantities of sulfur, phenols, and oil entrained in it.

Naphtha reforming.   The naphtha from the fractioning tower is subject to hydrogen treating, primarily to remove sulfur, and then goes to the catalytic reformer, the principal product of which is a high-octane gasoline-blending stock referred to as reformate.[17] The reformer also produces hydrogen, light hydrocarbon gases (C_1–C_3), butanes (C_4's), and a small amount of polymer, a mixture of relatively heavy and complex aromatic hydrocarbons. As noted above, these separate products must be obtained by fractional distillation of the product stream from the reformer. The necessity for such a separation step after nearly every reaction process adds considerably to the gross cooling-water use and waste-heat generation of the refinery. (See table 2 for a summary of products, utilities, residuals, and costs per barrel of input.) The reformer may be

[15] Mercaptans are chemical compounds having the general formula RSH where R is a radical (for example, R could be CH_3CH_2) and sulfur is present in its reduced state, as in hydrogen sulfide. Many mercaptans have extremely offensive odors. More advanced processes are available to do this same job, but there did not appear to be sufficient published information to include any of them in the model.

[16] There are actually two octane-rating scales with determinations by different tests. I use the Research Octane Number (RON) exclusively. Present premium gasolines have an RON $\geq$ 100, and present regular gasolines are 94–96 octane on this scale. Some of the new no-lead, low-octane gasolines are discussed later. The other scale is the Motor Octane Number. Motor octane is always lower than research octane and the spread is a measure of stability.

[17] Later in this chapter there is a discussion of hydrogen treating. Reforming is a transformation process in which the hydrocarbon molecules are rearranged to give a higher-octane product than the straight-run naphtha input. The net effect is to increase the number of carbon-to-carbon bonds and release hydrogen previously tied up in the naphtha.

TABLE 2.   Catalytic-Reforming Process: Product Yields, Utilities, Residuals, and Costs

Product yield/bbl input	Low-severity process		High-severity process	
	East Texas crude	Arabian mix	East Texas crude	Arabian mix
Hydrogen (*lb*)	5.18	5.20	5.72	5.70
C₃ and lighter (*lb*)	25.3	36.1	30.0	40.8
C₄'s—butane, butene (*lb*)	16.0	22.0	15.8	21.8
400°F end-point reformate (*bbl*)	0.80	0.75	0.77	0.73
Polymer (*bbl*)	0.035	0.019	0.035	0.022
Research Octane No. of reformate (clear)	97		100	
Utilities + residuals/bbl input	**High capital**		**Low capital**	
Gross fresh heat input (*10⁶ Btu*)	0.304		0.350	
Steam (*lb*)ᵃ				
E. Tex. crude	60		60	
Arabian mix	66.5		66.5	
Residual heat (*10⁶ Btu*)ᵇ	0.241		0.263	
Costs/bbl input	**Low-severity process**		**High-severity process**	
Low capital (*$*)	0.485		0.502	
High capital (*$*)	0.510		0.527	

ᵃ Condensate is assumed to be free of significant residuals.

ᵇ The heat of reaction is assumed to be 68,000 Btu/bbl of feed.

run using a high or low relative capital input and under more or less "severe" conditions of temperature and pressure in the reaction vessel. More severe conditions produce a smaller volume of higher-octane reformate at some additional cost (particularly in terms of spent catalyst).[18]

Hydrogen from the reformer goes to help meet the demand for this gas in the various hydrogen treating units. The reformer gases, being essentially free of sulfur, may either be sold as petrochemical feedstocks or burned to meet the refinery heat requirements, depending on relative prices.[19] Butanes (the C₄'s) from the reformer go either to the isomeriza-

[18] "More severe" does not mean higher in this case. Lower temperatures and pressures apparently result in reactions that are more difficult to control. The reformer is not quite current in this regard; my implicitly assumed pressure is about 400 psig; current practice is to use 200 psig, producing higher octanes and better yields. This difference will obviously affect the ability of the refinery in this study to make no-lead, high-octane gasoline. (Bechtel technical review.)

[19] Other refinery gases will generally have some hydrogen sulfide content and, as described below, must be desulfurized before they are salable for petrochemical feed. However, an option is included for the burning of these gases before desulfurization.

tion unit or to the alkylation unit. The isomerization unit is another transformation unit; it rearranges normal butane (nC_4H_{10}) to form the isomer, isobutane (iC_4H_{10}), with exactly the same number of carbon and hydrogen atoms but a slightly different shape and hence slightly different physical and chemical properties.[20] Isobutane in turn serves as one of the feeds to the alkylation unit.

To convert about 200 pounds (lb) of normal butane to isobutane requires about 180 lb of steam, generates 0.130×10^6 Btu of residual heat, and costs \$0.58. It is assumed that the condensate from the stripping and heating steam contains no significant quantities of residuals.[21]

Actually, isomerization units are seldom operated where they exist and are not currently being installed in new refineries, because various technological changes have made other sources of isobutane, both internal and purchased, cheaper. In particular, it will be seen that isobutane may be obtained internally from the catalytic cracker and from the new hydrogen-intensive cracking units. The isomerization unit is retained primarily to check the model's ability to reject its use.[22]

Finally, the polymer, a mixture of heavier hydrocarbons with a high proportion of aromatics,[23] may be sold as petrochemical feed or burned as refinery fuel.

Straight-run kerosene. The kerosene from the original distillation is sold, either as is or after being subjected to hydrogen treating to achieve desulfurization, in which case it is assumed to be acceptable as jet fuel.

Catalytic cracking: virgin gas oil. The next fraction, the combined light and heavy gas oils (the fraction of the crude boiling at between about 500°F and 950°F) is sent to a catalytic cracking unit, whose major function is to break the relatively long carbon chains in the gas oil to form shorter chains, especially those in the gasoline range. The by-products of this reaction are (1) a small amount of hydrocarbon gas (handled as de-

---

[20] In general, isomers have slightly higher octane ratings than their normal counterparts.

[21] See Beychok, *Aqueous Wastes*, p. 37.

[22] There is a prospect that isomerization of C_5–C_6 streams will be useful if refiners have to go to production of high-octane, low- or no-lead gasoline. See J. C. Dunmyer, Jr., R. E. Froelich, and J. L. Putnam, "The Cost of Replacing Leaded Octanes" (paper presented at 36th Midyear Meeting, API, Division of Refining, San Francisco, 13 May 1971).

[23] Aromatics are compounds with one or more carbon ring structures. Benzene, toluene, and xylene are the aromatics of interest in the polymer. A mixture of these three aromatics is often referred to as BTX.

scribed below); (2) a C_4 or butane-butene "cut," which may go either to alkylation or to refinery gases; (3) some isobutane; (4) an amount of relatively heavy stock that is called cycle oil and may be thought of as the uncracked portion of the input gas oil; and (5) petroleum coke, which is deposited on the catalyst. Yields, utilities, residuals, and costs are summarized in table 3. This table reflects the assumption that new zeolite

TABLE 3. Fluid Catalytic Cracking of Virgin Gas Oil: Yields, Utilities, Residuals, and Costs

	East Texas crude[a]		Arabian crude[b]	
	Low conversion	High conversion	Low conversion	High conversion
Yield/bbl of fresh feed[c]				
Gases: C_1–C_3 (*lb*)	19.1	30.5	25.0	35.5
Sulfur in gas (%)[d]	1.41	1.04	5.11	4.20
C_4 { Butane, butene (*lb*)	17.2	25.9	17.2	25.9
C_4 { Isobutane (*lb*)	8.6	15.2	8.6	15.2
Cat gasoline (/*bbl* @ *270 lb/bbl*)	0.532	0.545	0.513	0.580
Gasoline S weight before debutanization (%)[e]	0.081	0.04	0.4	0.2
Gasoline RON after caustic scrubbing	95		90	
Cycle stock (*bbl*)	0.331	0.165	0.331	0.165
Coke (*lb*)	18.0	26.0	18.0	26.0
Utilities and residuals/bbl fresh feed[f]				
Fresh heat input (*10^6 Btu*)	0.0985	0.160	0.0985	0.160
Boiler water intake (*gal*)	8.88	15.71	8.88	15.71
Net steam production (*lb*)	39	56	39	56
Residual heat (*10^6 Btu*)	0.155	0.258	0.155	0.258
Condensate (*gal*)	4.2	9.0	4.2	9.0
Flue gas from regenerator (*10^3 acf*)	4.18	6.04	4.18	6.04
SO_2 (*lb/bbl*)	0.30	0.56	1.44	2.68
Particulates (*lb/bbl*)	0.22	0.32	0.22	0.32
Cost/bbl of fresh feed ($)	0.2920	0.4047	0.2920	0.4047

[a] Feedstock weight/bbl = 309 lb.   Sulfur weight = 0.36%.
[b] Feedstock weight/bbl = 310 lb.   Sulfur weight = 1.65%.
[c] Based on D. H. Stormont, "Texaco Fluid Catalytic Cracker Incorporates Novel Features," *Oil and Gas Journal*, vol. 66, no. 14 (April 1968), pp. 114–15. Modified in accordance with the Bechtel Corporation's technical review of an earlier draft of this study, 12 April 1972. "Fluid" refers to the method by which the catalyst is moved back and forth from reaction to regeneration vessels.
[d] After allowing for S (as H_2S) dissolved in condensate.
[e] After debutanization, sulfur in the gasoline is assumed to be 0.005% for all columns.
[f] Utilities and volume of condensate were previously estimated for cat crackers using the old catalysts. I simply assume these are the same for the new catalysts and have made a similar assumption for costs. These assumptions, however accurate or inaccurate, probably have no significant effect on the results for the basic model, since there is no real choice of processing paths for gas oil. Thus what is important is the relation between the two recycle alternatives; I have no reason to think that the present estimates are inaccurate for this comparison.

catalysts are used in the cracking unit. These catalysts allow higher conversion rates and greater gasoline yield, while producing less coke and gas than the older catalysts. They have been responsible for a recent resurgence in cat cracker building activity. For the same catalyst, higher conversion rates and greater gasoline yields are obtained by recycling all or part of the cycle oil. In the process more gas and coke are also produced.[24]

The cat gasoline, after the removal of butanes, is scrubbed with caustic to remove the high phenols concentrations resulting from the severe conditions in the cracking vessel. It then becomes part of the gasoline blending pool. The gases may be burned directly for refinery fuel or subjected to desulfurization, after which they may be mixed with the reformer gases for sale as petrochemical feed. The isobutane is sent to the alkylation unit, while the other butanes may be burned, combined with the other refinery gases, or also sent to the alkylation unit. The cycle stock may be sold as distillate fuel oil or used as cutter stock with the reduced crude from the vacuum tower. The resulting mixture of reduced crude and cycle oil is sold as residual fuel oil, either before or after desulfurization through hydrogen refining.

The coke, essentially pure carbon, deposited on the catalyst in the reaction chamber interferes with catalyst activity and must be burned off. Indeed modern fluid catalytic crackers may involve catalyst circulation from reaction chamber to regeneration vessel on the order of 2 tons/bbl of fresh feed.[25] In the process of regeneration large quantities of heat are generated, some of which is captured and used in the process, largely by mixing the feedstock with the hot, freshly regenerated catalyst before entry into the reaction chamber, but also by making steam both for internal use and for "export" to the rest of the refinery. Regeneration also

---

[24] For example, consider a simplified process in which 1 lb of input is converted to 0.5 lb of cycle oil, 0.4 lb of gasoline, and 0.1 lb of "other." If a similar unit were built to take the cycle oil, for every pound of original feed, the second unit would produce 0.2 lb of gasoline, 0.05 lb of "other," and 0.25 lb of cycle oil. Looking at the two processes together, the yields per pound of fresh feed would be 0.25 lb of cycle stock, 0.6 lb of gasoline, and 0.15 lb of "other." The costs involved in obtaining this extra yield are those of the additional capacity and utilities required to deal with the cycle oil from the first unit. If the two units are in fact collapsed into one, with a ratio of recycle to fresh feed of 0.5, the same results are obtained. In the model developed here, overall cat cracker capacity is variable, and the costs of obtaining higher yields are reflected in the utilities (and residuals) per barrel of fresh feed and in the objective function values that reflect capital costs for building larger capacity.

[25] The average from a survey of fluid catalytic cracking units at Los Angeles refineries is reported in U.S. Public Health Service, *Atmospheric Emissions from Petroleum Refineries.*

produces large quantities of airborne particulates (soot and catalyst fines) and a quantity of SO_2, depending on the quantity of sulfur in the coke. Most of the particulate matter is trapped in the regeneration vessel as a matter of standard practice because of the value of the catalyst that otherwise would be lost. In chapter 4 alternatives are discussed for dealing with the untrapped portion. In accordance with the definition given in the introduction, only this untrapped portion is included in residuals generation in the model.

Coking of reduced crude. The reduced crude yielded by distillation may be sold as residual fuel oil (with or without desulfurization) or burned to meet the refinery's fuel needs; or it may be routed to the coker, where again the long chains are broken into shorter ones, thereby increasing the yield per barrel of crude of the lighter hydrocarbons. The products of coking are light hydrocarbon gases, gasoline, gas oil, and coke. Table 4 shows the product distribution per barrel of feed, along with utilities requirements, residuals generation, and costs. The condensate from this process has one of the highest residual loads in the refinery (see chapter 4). Note that coking is also provided with the capital versus fresh and waste heat trade-off discussed above.

The coker gases are handled as described above for the catalytic-crack-

TABLE 4.   Coking of Reduced Crude: Product Yields, Utilities, Residuals, and Costs

Product yields/bbl feed	East Texas[a] crude	Arabian mix[b] crude
Gases, C_1–C_3 (*lb*)	23.7	25.8
Sulfur in gas (%)[c]	1.85	3.40
Coker gasoline (*bbl*)	0.253	0.250
Coker gas oil (*bbl*)	0.557	0.560
Coke (*lb*)	70.3	82.1
Sulfur in coke (%)	1.98	3.32
Utilities and residuals/bbl feed	Low capital	High capital
Fresh heat input (*10⁶ Btu*)	0.279	0.242
Boiler water intake (*gal*)	17.4	17.4
Net steam production (*lb*)	20	20
Residual heat (*10⁶ Btu*)	0.269	0.251
Condensate (*gal*)	15.0	15.0
Cost/bbl input (*$*)	0.264	0.298

Note: Yields, costs, etc., are based on W. L. Nelson, *Guide to Refinery Operating Costs* (Tulsa, Okla.: Petroleum Publications, 1970), p. 65; and Bechtel Corporation's technical review of an earlier draft of this study, 12 April 1972.

[a] Feedstock weight/bbl = 338 lb.   Sulfur weight = 1.28%.
[b] Feedstock weight/bbl = 345 lb.   Sulfur weight = 3.33%.
[c] After allowing for sulfur dissolved (as H_2S) in the condensate.

TABLE 5. Catalytic Cracking of Coker Gas Oil: Yields, Utilities, Residuals, and Costs

	East Texas crude[a]		Arabian mix crude[b]	
	Low conversion	High conversion	Low conversion	High conversion
Yield/bbl fresh feed				
Gases: C_1–C_3 (*lb*)	20.1	32.2	20.3	32.5
Sulfur in gas (%)[c]	4.61	3.14	12.2	9.15
C_4 {Butane, butene (*lb*)	18.0	27.2	18.2	27.9
{Isobutane (*lb*)	9.1	16.0	9.2	16.0
Cat gasoline (*bbl*)	0.528	0.584	0.535	0.590
Gasoline S weight before debutanization (%)	0.26	0.14	0.76	0.40
Gasoline RON after caustic scrubbing	90		87	
Cycle stock (*bbl*)	0.331	0.165	0.331	0.165
Coke (*lb*)	19.2	28.0	19.2	28.8
Utilities and residuals/bbl fresh feed				
Fresh heat input (*10^6 Btu*)	0.0985	0.160	0.0985	0.160
Net steam production (*lb*)	41	60	42	62
Residual heat (*10^6 Btu*)	0.155	0.258	0.155	0.258
Condensate (*gal*)	4.2	9.0	4.2	9.0
Flue gas from regenerator (*10^3 acf*)	4.47	6.60	4.47	6.80
SO_2 (*lb*)	1.28	1.88	3.10	4.44
Particulates (*lb*)	0.24	0.35	0.24	0.36
Hydrocarbons (*lb*)	0.218	0.338	0.218	0.338
Cost/bbl of fresh feed (*$*)	0.2920	0.4047	0.2920	0.4047

[a] Feedstock weight/bbl = 316 lb.   Sulfur weight = 1.20%.
[b] Feedstock weight/bbl = 320 lb.   Sulfur weight = 3.50%.
[c] After allowing for sulfur as H_2S dissolved in condensate.

ing unit. The coker gasoline is sent to a hydrotreater for removal of sulfur and phenols, and the resulting stream is then subject to catalytic reforming in order to improve its octane rating.[26] Coker gas oil is catalytically cracked to produce more gasoline-blending stock. The yields from this process are slightly different from those resulting from cracking virgin gas oil. More significant differences, however, are to be observed in the process residuals, because of the higher sulfur and phenol content of the coker gas oil.[27] The yields, utilities, residuals (except for those in the foul condensate), and costs are shown in table 5.

[26] In principle, coker gasoline could also be sent to gasoline blending either directly or after hydrotreating. The former step is most unlikely because of the relatively high sulfur and phenol content of this stream. I have not included the latter possibility either, primarily because I could find no information on which to base estimates of the octane numbers for these streams.

[27] The higher sulfur content in the coker gas oil results from the relatively large amount of the original sulfur that remains in the reduced crude subject to coking. The higher phenol content is an essentially uncontrollable side effect of the high pressure and temperature required for coking. Under these severe conditions, some phenols are inevitably formed.

The petroleum coke, consisting of nearly pure carbon, may be sold or used as fuel within the refinery. The price obtained will depend on the sulfur content—sweet (low-sulfur) cokes selling for considerably more than sour (high-sulfur) grades.

Alkylation. The alkylation unit combines the butane-butene stream from the cat cracker and reformer with isobutane (purchased or provided internally from the cat cracker or the isomerization unit) to produce iso-octane (iC_8H_{18}) or alkylate, a very high-octane gasoline-blending stock. This reaction is catalyzed by sulfuric acid and is strongly exothermic, hence the major residuals are spent sulfuric acid and waste heat. A large amount of steam is required, but it is assumed that the condensate is clean.[28] The inputs and outputs can be summarized as follows:[29]

Input	isobutane	57 lb
	butane-butene	215 lb
Output: alkylate		1 bbl
Utilities: steam		290 lb
Residuals	spent acid	15.01 lb
	heat	0.6×10^6 Btu
Research Octane No. of alkylate		99
Cost per barrel of alkylate		$1.50

Gasoline blending. In the gasoline-blending section of the refinery, the separate gasoline-blending stocks are combined to make regular and premium gasolines for sale. To each stock may be added 1, 2, or 3 cubic centimeters of TEL per gallon to improve the octane rating of the stock. Table 6 shows the 11 blending stocks potentially available in the model and summarizes their clear octane numbers as well as their response to the addition of lead.[30] It is assumed that the cracked gasolines must be caustic scrubbed to remove high phenol concentrations, but that straight-run gasolines may or may not be scrubbed. The reformates and alkylate are quite "clean" without scrubbing because of the requirements that the feeds to these processes be relatively free of sulfur and other impurities.

The gasoline blends in the basic model must meet certain requirements: minimum octane of regular = 94, of premium = 100; maximum

[28] Based on a description of effluent from alkylation units in Beychok, *Aqueous Wastes*, p. 36.
[29] Inputs and utilities are based on the Bechtel technical review; the cost on Nelson, *Guide to Operating Costs*, p. 89.
[30] These numbers are based primarily on estimates provided by the Bechtel technical review, adjusted to try to make the catalytically produced stocks (reformate, cat cracker gasoline) more representative of these products generally.

TABLE 6. Gasoline–Blending Stocks: Octane and Lead Susceptibility

	Reformate: low-severity process	Reformate: high-severity process	Alkylate	Straight run: E. Texas	Straight run: Arabian	Scrubbed straight run: E. Texas	Scrubbed straight run: Arabian	Cat gasoline: E. Texas virgin gas oil (scrubbed)	Cat gasoline: Arabian virgin gas oil (scrubbed)	Cat gasoline: E. Texas coker gas oil (scrubbed)	Cat gasoline: Arabian coker gas oil (scrubbed)
Research Octane No., clear	97.0	100.0	99.0	70.0	58.0	70.0	58.0	95.0	90.0	90.0	87.0
RON with 1 cc TEL	99.4	102.3	103.5	75.0	63.0	76.5	65.0	96.5	92.2	94.0	92.6
RON with 2 cc TEL	101.0	103.5	106.0	78.9	66.0	80.5	68.0	97.8	93.6	96.3	95.6
RON with 3 cc TEL	102.0	104.5	108.0	81.6	68.0	83.0	70.0	98.9	95.0	98.0	97.0

lead content = 2.5 cubic centimeters per gallon.[31] The general method
of introducing these requirements was described in chapter 2. In the
specific case of the octane requirements, if there are n stocks with octane
levels $O_1, \ldots, O_n$ and it is required that the blend resulting from these
stocks have an octane level of at least $\bar{O}$, it is assumed that the octane of
the blend can be approximated by the volume-weighted average of the
individual octanes.[32] Then if the volumes blended are $Q_1, \ldots, Q_n$, the
octane constraint may be written:

$$\frac{Q_1 O_1 + Q_n O_n}{Q_1 + \ldots + Q_n} \geq \bar{O},$$

from which the following is obtained:

$$Q_1(O_1 - \bar{O}) + \ldots + Q_n(O_n - \bar{O}) \geq 0.$$

Thus in the columns representing the unit-blending activities, the octane
constraint row entries are $O_1 - \bar{O}$, and so on, and the right-hand is zero.
The same method is applied in the model to the maximum lead allowable.
No simplification beyond that already noted is involved in the application since the method involves taking the volume-weighted average of
cubic centimeters per unit volume.

"Stability" has already been mentioned as another characteristic of
blended gasoline. Ideally the model would include lower limits on stability and upper limits on Reid Vapor Pressure (essentially a measure of how
fast the gasoline will evaporate at ambient temperature) as well as those
on octane and lead.

Hydrogen Treating for Impurity Removal

Probably the most important development in refining in the last decade has been the introduction of a number of processes in which excess

[31] This is said to be the level currently permitted by an informal government-
industry agreement, but I have found no published evidence of this. (Compare, however, the use of a 3 cc/gal TEL limit for current gasolines by Dunmyer, Froelich, and
Putnam, "Cost of Replacing Leaded Octanes.") The octane and lead requirements
are two of the key parameters experimented with below in studying alternative future
patterns of water use and residuals generation. No limits are placed on the sulfur and
phenol content in the basic model because no such limits are published and accepted.
See American Society for Testing Materials, *1969 Book of ASTM Standards*, pt. 17
(Philadelphia, 2d ed., 1969), pp. 184–296. Such standards are published for aviation
gasoline, ibid., p. 296.
[32] This assumption about the behavior of octane ratings of blends is what makes
the application possible in this case. While it is not strictly accurate, it does seem safe
to assume that no systematic bias is introduced by making it.

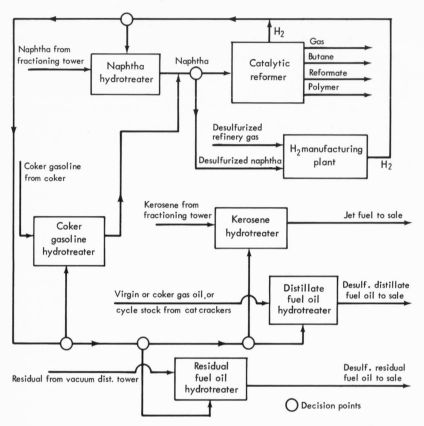

Figure 8. The hydrogen generation and use system. The process units in this figure are the ones that appear in the basic refinery. Some newer H_2-using processes are described in the text.

hydrogen is introduced to the reaction vessel. These units appear in fig. 8. Depending on the amount of hydrogen provided per unit of the input, and on other conditions maintained in the vessel, these so-called hydro-processes may be used to reduce impurities (hydrotreating), to effect molecular changes in a relatively minor part of the input (hydrorefining), or to reduce the molecular size of a major portion of the input (hydro-cracking).[33] The refinery described so far contains examples of the first and second of these process types, but the hydrocracking processes are brought in below as important components of a refinery in the 1970s in order to exhibit the effects on residuals of technological changes.

Hydrotreating is primarily directed at the removal of sulfur from such

[33] This division by function and the accompanying nomenclature was suggested by D. H. Stormont in "Nomenclature for Hydroprocessing," *Oil and Gas Journal*, vol. 66, no. 41 (7 October 1968), pp. 174–75.

refinery streams as the straight-run naphtha going to the reformer and kerosene that is being upgraded to meet jet fuel quality requirements. The excess of hydrogen encourages the extraction of both elemental and chemically bound sulfur and produces, along with the purified product, a quantity of off-gas rich in hydrogen sulfide and a condensate in which some of that H_2S is dissolved. The sulfur removal percentages attainable with this process are impressive—95 percent and over—and all the removed sulfur becomes a residuals problem, ultimately being discharged as SO_2 from the refinery stacks or as H_2S from its sewer outfalls, *unless* sulfur recovery, as described in chapter 4, is practiced.

Stormont suggests that "hydrotreating" be used to refer to processes using less hydrogen than 100 standard cubic feet (scf)/bbl of input stream and proposes the label "hydrorefining" for processes using between 100–1,000 scf/bbl of input.[34] The latter more hydrogen-intensive processes also have sulfur removal as a prime goal, but for heavier fractions the amount of sulfur involved may require considerably more than 100 scf/bbl of hydrogen for removal. In addition the heavier fractions, as well as certain light products of cracking reactions, are liable to contain significant amounts of aromatic compounds, such as the phenols, characterized by the presence of carbon rings and generally lending undesirable combustion or reaction properties to the stream. Thus, for example, coker gasoline may be high in aromatics because of the severity (heat and pressure) of the conditions surrounding coking. Before this gasoline is reformed to upgrade its octane rating, it must be hydrotreated to remove not only sulfur, but also aromatics. Often the desulfurization of residual fuel oil will fit under this hydrorefining label. As an additional example some refineries presently hydrorefine the virgin and coker gas oils before sending them through the cat-cracking process because this results in better cracking yields and longer catalyst life.[35]

In the basic refinery model, hydrogen-treating or hydrogen-refining processes are provided for several streams: straight-run naphtha and coker gasoline must be hydrotreated before reforming; straight-run kerosene may be hydrotreated for sulfur and aromatics removal and sold as jet fuel;[36] cat cracker cycle oil destined for sale as distillate fuel oil may

[34] Ibid.
[35] See, for example, W. F. Arey, Jr., and L. Kronnenberger, "Hydrotreating Cat-cracker Feedstocks for Attractive Economic Benefits," *Oil and Gas Journal*, vol. 67, no. 20 (19 May 1969), pp. 131–39.
[36] See R. C. Hansford and others, "Unisar's New Hydrogeneration Process Saturates Aromatics in Jet Fuel," *Oil and Gas Journal*, vol. 67, no. 18 (5 May 1969), pp. 134–36.

be hydrotreated and sold as low-sulfur fuel oil; and finally, residual fuel oil may be desulfurized for sale as low-sulfur residual to allow customers to meet air quality control standards. The same methods were employed in estimating unit costs, fuel, steam and water requirements, and hydrogen consumption for each of these processes, so it will be convenient simply to describe the methods once and present the resulting process vectors in tabular form.

The basic parameter to be estimated for each process is the hydrogen consumption per barrel of feed. This is a function primarily of the amount of sulfur to be removed, which in turn depends on the sulfur in the input stream and the desired sulfur in the output.[37] Estimates of the fate, at each previous step, of the sulfur in the original crude provide a figure for the sulfur concentration in each of the feedstocks. The desired level of sulfur in outputs varies with the fraction involved, as table 7 points out.

Once hydrogen consumption has been determined, utilities and costs may both be interpolated from information provided by Nelson for processes using 300, 700, 1,000, and 1,100 scf of hydrogen/bbl of feed.[38] The costs given for fuel, steam, and cooling water may be translated back into physical terms to fit into the framework of the model. The resulting process vectors are presented in table 8.

For the basic refinery, with a moderate amount of hydrotreating being accomplished, a sufficient supply of hydrogen can be recovered from the catalytic reformer gases. When the amount of hydrotreating is large, however, or hydrocracking processes are used to break up long-chain hydrocarbons, this source must be supplemented. For this purpose it is becoming standard for a refinery to be equipped with a hydrogen-manufacturing plant that catalytically reforms steam and one or another of the light hydrocarbons into hydrogen. The refinery model has a hydrogen plant option for which the alternative feedstocks are desulfurized refinery gas and desulfurized naphtha.[39] The relation between hydrocarbon feedstock and steam was determined by satisfying two requirements:

1. the sum of the hydrogen weight in the steam and in hydrocarbon input must be 3 lb for every pound of hydrogen output; and

2. the molar ratio of steam to carbon inputs must be maintained at

---

[37] For relations among feedstock sulfur content, product sulfur content, and hydrogen consumption, see W. L. Nelson, "Hydrogen Needed for Residual Desulfurization," *Oil and Gas Journal*, vol. 67, no. 48 (24 November 1969), pp. 79–80.

[38] W. L. Nelson, "Operating Costs," *Oil and Gas Journal*, vol. 66, no. 38 (16 September 1968), pp. 96–97; idem, "Residual Desulfurization Costs," ibid., vol. 66, no. 48 (25 November 1968), p. 131; and idem, "Process Costimating 73-A: Operating Costs

TABLE 7.   Hydrogen Requirements for Desulfurization of Various Feedstocks

	Sulfur in feed (percent)	Sulfur in output[a] (percent)	H₂ required per barrel[b] (scf)
Naphtha to reformer			
East Texas crude	0.1	0.005	100
Arabian crude	0.4	0.005	205
Coker gasoline to reformer			
East Texas crude	0.2	0.005	250
Arabian crude	0.5	0.005	480
Kerosene to jet fuel			
East Texas crude	0.15	0.1	220
Arabian crude	0.70	0.1	565
High-sulfur distillate fuel oil[c]	3.80	0.3	1,310
Medium-sulfur distillate fuel oil[c]	2.80	0.3	1,040
Residual 1[d]	1.18	0.5	400
Residual 2[d]	1.29	0.5	420
Residual 3[d]	2.81	0.5	1,050
Residual 4[d]	3.01	0.5	1,080

[a] The output sulfur requirements were determined on a variety of bases. The reformer feedstock limits are more or less arbitrary but reflect the sensitivity of the process catalyst to small amounts of impurities. The jet fuel limits are based on the proposed specifications for military jet fuel JP-7 in A. V. Churchill, "Characteristics of Advanced Military Turbine Fuels" (paper presented at the National Petroleum Refiners Association Annual Meeting, San Antonio, Texas, 23–26 March 1969). These requirements are more stringent than for earlier military fuels and for civilian standards set by the American Society for Testing Materials. The limit of distillate fuel oil was chosen to match the sulfur content of currently sold No. 2 fuel oil (according to the Bechtel Corporation's technical review of an earlier draft of this study, 12 April 1972). For residual fuel oil, the limit is near the low end of the scale of commercially available sulfur contents and corresponds to limits set in several areas to cut down on SO₂ emissions.

[b] Hydrogen requirements were determined by applying and extrapolating relations provided in W. L. Nelson, "Hydrogen Needed for Residual Desulfurization," *Oil and Gas Journal*, vol. 67, no. 48 (24 November 1969), pp. 79–80. Modified according to advice in the Bechtel technical review.

[c] High sulfur distillate fuel oil is the cycle stock from the cat cracking of coker gas oil from Arabian crude. Medium-sulfur distillate fuel oil may be either of the following: virgin gas oil, cat cracker cycle stock from Arabian crude; or coker gas oil, cat cracker cycle stock from East Texas crude.

[d] Residual fuel oil is a mix of the reduced crude from the vacuum tower and a "cutter" oil used to reduce viscosity. I use cat cracker cycle oil for this cutter in the proportion of 1 to 10 with the residual. The four grades of residual fuel oil included represent the following combinations of sulfur-content oils:

	Reduced crude	Cutter stock (distillate fuel oil)
Residual 1	low sulfur (East Texas crude)	low sulfur
Residual 2	low sulfur (East Texas crude)	medium sulfur
Residual 3	high sulfur (Arabian crude)	medium sulfur
Residual 4	high sulfur (Arabian crude)	high sulfur

of Hydrogen Treating," ibid., vol. 11, no. 9 (1 March 1971), pp. 64–65.

[39] The process vectors are based on data in J. Voogd, "Hydrogen Plants Show Good Reliability and Low Costs," *Oil and Gas International*, vol. 11, no. 9 (September 1971), pp. 77–80.

TABLE 8. Hydrotreating Activities: Per Barrel Quantities

	Product vol. (bbl)	Off-gases, 25% H_2S (lb)	Hydrogen[a] (lb)	Fresh heat (10^6 Btu)	Residual heat (10^6 Btu)	Steam (lb)	Condensate (gal)	Unit cost (net of hydrogen, steam, heat, cooling water) ($)
Straight-run naphtha								
East Texas	1.00	0.85	0.53	0.04	0.046	5.5	0.66	0.259
Arabian mix	0.996	3.06	1.09	0.04	0.046	10.5	1.36	0.269
Coker gasoline								
East Texas	1.00	1.36	1.32	0.06	0.066	7.0	0.84	0.311
Arabian mix	1.00	3.44	2.54	0.06	0.066	14.0	1.68	0.332
Straight-run kerosene								
East Texas	1.00	0.09	1.16	0.08	0.086	10.0	1.2	0.271
Arabian mix	0.987	4.84	3.00	0.08	0.086	20.0	2.4	0.310
Medium sulfur distillate fuel oil	0.964	22.2	5.50	0.075	0.081	32.6	3.9	0.354
High sulfur distillate fuel oil	0.942	33.2	6.95	0.080	0.086	35.0	4.2	0.460
Residual fuel oil 1[b]	1.00	3.86	2.12	0.125	0.131	25.0	3.0	0.267
Residual fuel oil 2[b]	0.995	5.31	2.22	0.125	0.131	25.0	3.0	0.270
Residual fuel oil 3[b]	0.965	15.7	5.56	0.112	0.118	41.5	5.0	0.342
Residual fuel oil 4[b]	0.978	17.5	5.72	0.112	0.118	41.5	5.0	0.346

[a] One pound of hydrogen at standard conditions of temperature and pressure—60°F and 14.7 psi, absolute—(that is, atmospheric pressure at sea level) is equivalent to 189.5 cubic feet. Hence the conversion from volume in table 7 to weight in this table.

[b] Quantities are expressed per barrel of total input, i.e., roughly 0.9 bbl of reduced crude plus 0.1 bbl of cycle oil cutting stock.

3.8 for both feedstocks [40] (this is to prevent carbon coking on the catalyst and consequent plant shutdown for cleaning).

TABLE 9. Hydrogen-Producing Plant: Inputs, Utilities, Residuals, and Unit Costs

	Desulfurized refinery gas feed	Desulfurized naphtha feed
Refinery gas (*lb*)	4.7	—
Naphtha (*lb*)	—	4.35
Boiler feed water for steam (*lb*)	17.9	20.6
Fresh heat input (10^6 *Btu*)	0.0833	0.0958
Net steam production (*lb*)	0.0122	0.0140
Waste heat (10^6 *Btu*)	0.0064	0.0074
Cost: capital + operation (*$*)	0.0147	0.0165

Note: These figures are for a plant designed to produce 318,000 lb (60×10^6 scf) per day of hydrogen. All entries are per pound of H_2 produced.

Table 9 shows the inputs, utilities, residuals, and unit costs for a plant designed to produce 318,000 lb (60×10^6 scf) per day of hydrogen.

Sour Gas Desulfurization

The gases (C_1–C_3) from the coker, catalytic cracker, and hydrotreaters are "sour"; that is, they have a significant amount of hydrogen sulfide entrained. If such gas is to be sold for petrochemical feed, this H_2S must be removed, and the model includes a gas desulfurization option for this purpose.[41] (Figure 9 shows the gas collection system for the refinery, including units producing gases free of H_2S.) Any number of specific processes may be used to accomplish desulfurization; the activity in the model does not follow any single one of these but is based on composite information supplied by Nelson.[42] The basic assumption is that 96 percent of the H_2S in the gas is recovered; of the unrecovered portion, half is assumed to remain in the gas and half to be lost to the atmosphere.[43] In addition the process generates waste heat and requires inputs of refinery

[40] Ibid., p. 78. This simply says that the weight of steam divided by its mole weight (18) must be 3.8 times as great as the weight of carbon in the gas or naphtha divided by 12 (the mole weight of carbon). Since for different hydrocarbons, carbon is a different fraction of total weight, the ratio of steam to hydrocarbon will vary with the feedstock.

[41] This option also applies to the sour gases generated in the catalytic cracking of coker gas oil.

[42] See W. L. Nelson, *Guide to Operating Costs*, pp. 103–4; idem, "Cost of Plants for Recovering Hydrogen Sulfide," *Oil and Gas Journal*, vol. 66, no. 17 (22 April 1968), p. 199; and "What Does It Cost to Operate Gas-Desulfurization Plants?" ibid., vol. 66, no. 32 (5 August 1968), p. 125.

[43] For convenience it is assumed that this gas has been oxidized to SO_2 before escaping.

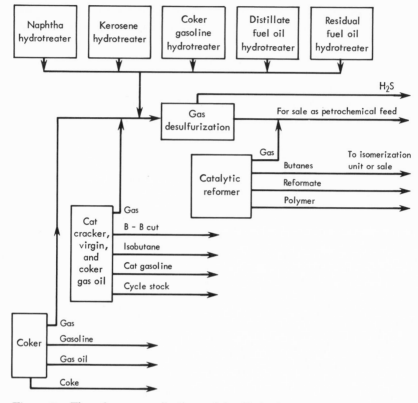

Figure 9. The refinery gas collection and desulfurization system.

steam, both in amounts varying nonlinearly with the sulfur content of the gas. In the model a separate activity is provided for each sour-gas stream in order to reflect this variation. As an example of the resulting set of vectors, consider the one for desulfurization of cat cracker gas resulting from the cracking to 66.9 percent conversion of virgin East Texas gas oil. The unit basis for the vector is 80.5 lb of sour gas (1,000 scf for a gas with an average molecular weight of 30.5). The activity values are:

Sour gas in	80.5 lb
H_2S content of gas in	1.5 %
Refinery gas out	79.3 lb
H_2S content of gas out	0.031 %
H_2S recovered	1.16 lb
SO_2 to atmosphere	0.047 lb
Steam required	6 lb
Residual heat load	0.0038×10^6 Btu
Cost per 1,000 scf of sour gas input	$0.022

The recovered H_2S may then be burned (flared) producing SO_2 emissions, or it may be sent to a sulfur recovery plant. (The latter is discussed in chapter 4.)

TECHNOLOGICAL CHANGE IN REFINING PROCESSES

In order to explore the effect on residuals generation of changes in process technology, two versions of the refinery have been constructed. One contains only the processes described thus far; the other contains these processes plus four hydrogen-intensive cracking options. In the added processes, the reaction temperature and pressure and the quantity of excess hydrogen are such that a considerable fraction (over half for definitional purposes) of the feedstock molecules are reduced in size. The effect of applying this process to gas oil or reduced crude feed is generally to give a total yield volume greater than the input volume. One of the major advantages of the process is that the composition of the product stream can be varied within fairly wide limits. This applies particularly to the ability to concentrate on making either gasoline stocks or kerosene-fraction turbine fuels, especially jet fuels. It is possible to run a single hydrocracking unit in either way (though there is some cost to building in the flexibility), and hence a single refinery can be built with some assurance that its output composition can vary to meet a substantial range of possible future demand patterns. An additional advantage of hydrocracking is that hydrogen refining tends to proceed simultaneously with cracking so that the products will be low in sulfur and phenols and other undesirable constituents. The gases and condensate streams from these processes will be relatively high in sulfur and will contribute sig-nificantly to the quantity of by-product sulfur the refinery is capable of producing. (Sulfur recovery is discussed in chapter 4).

Figure 10 shows the place of the two hydrocracking processes in the context of the basic refinery described above.[44] The products of the cracking of reduced crude (H-oil process) are refinery gases, butanes, and so forth; a sulfur-free naphtha ready to go to the catalytic reformer, some kerosene, and a very heavy residue called "pitch," which is assumed

---

[44] The process as applied to the reduced crude is based on the H-oil process, espe-cially designed to take heavy fractions. See A. R. Johnson, S. B. Alpert, and L. M. Lehman, "Processing Residual Fractions with H-oil," *Oil and Gas Journal*, vol. 66, no. 26 (24 June 1968), pp. 86–94.

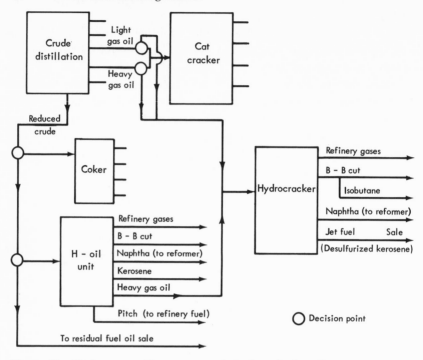

Figure 10. Hydrogen-cracking processes: their places in the refinery.

either to be burned for refinery fuel or disposed of for $0.50 per barrel.[45]
For each crude, a choice between two yield compositions is provided: one,
referred to as "high conversion," producing primarily kerosene and a
heavy gas oil (which may in turn be subject to hydrocracking or conventional cracking); and a second involving cycling the gas oil to extinction
and producing even more kerosene and some naphtha (reformer feed).

The other hydrocracking process is applied to the virgin gas oil from
either crude. Its products are refinery gases, a butane cut (and a separable
stream of isobutane), naphtha (bound again for the reformer), and kerosene (actually of jet fuel quality). Again the basic process may be operated to emphasize the production of kerosene or of naphtha, depending
on the pattern of demand.[46] The feed in the former case is heavy gas oil
and in the latter, light gas oil. The standard catalytic-cracking activities
are modified to feed a mix of heavy and light gas oils equivalent to their

[45] Actually, pitch probably has some value as a raw material for petrochemical production, but the assumption on disposal is conservative in the sense that it tends to
overstate the costs of reducing residuals discharges.
[46] For this modification I differentiate between the heavy and light gas oils from the
distillation process, something I did not do in the original model.

feed in the basic model. Excess supplies of either heavy or light gas oils are routed to distillate fuel oil production. The yields for the processes are summarized in table 10.

Costs and utilities for the several hydrocracking alternatives have been estimated in a manner basically similar to that described above for the desulfurization processes. First, the hydrogen consumption is determined for each process from the sources used to estimate process yields. Then, using this hydrogen consumption, costs and utilities are estimated using information published by Nelson.[47]

PRODUCTS: CONSTRAINTS AND PRICES

In order to assure a realistic product mix and comparability with other studies, the refinery model is constrained to produce at least minimum quantities of five important products. The constraints have been chosen to agree as closely as possible with those used by Dunmyer, Froelich, and Putnam in their study of the cost of restrictions on TEL additions.[48] The basic production constraint set is summarized below:

	Barrels per calendar day
Regular gasoline	51,150
Premium gasoline	35,100
Kerosene plus jet fuel	15,000
Distillate fuel oil (approximately No. 2 oil)	17,400
Residual fuel oil (approximately No. 6 oil)	3,000

A number of comments are in order on this approach and later modifications. First, there is no incentive for the model to produce more than these quantities, since no credit is given for production of these basic products. This means that the model is cast in the cost minimization mold. Earlier experience with a model set up to maximize profit, with prices on all products, demonstrated the great difficulty in obtaining and maintaining a realistic product mix in the course of changing TEL limits, other product quality measures, and so forth. Second, it is possible to vary the mix of product qualities (sulfur contents) within the categories of "kerosene plus jet fuel," "distillate fuel oil," and "residual fuel oil."

[47] Nelson, "Operating Costs," pp. 96–97. This information was supplemented from Johnson, Alpert, and Lehman, "Processing Residual Fractions."

[48] Dunmyer, Froelich, and Putnam, "Cost of Replacing Leaded Octanes." The mix was in turn based on U.S. Bureau of Mines figures for U.S. production in 1969.

TABLE 10. Yields of Hydrocracking Processes for Reduced Crude and Virgin Gas Oil

Yield[a] and cost	H-oil process				Hydrocracking virgin gas oil			
	East Texas		Arabian mix		East Texas		Arabian mix	
	High conv., some gas oil produced	Gas oil recycled to extinction	High conv.	Extinction	Maximum gasoline	Maximum jet fuel	Maximum gasoline	Maximum jet fuel
Refinery gases (lb)	16.2	21.9	16.6	22.4	12.4	5.3	12.4	4.8
Butanes, etc. (lb)	7.8	9.2	7.8	9.2	—	—	—	—
Normal butanes, etc. (lb)	—	—	—	—	12.7	9.3	12.7	9.3
Isobutane (lb)	—	—	—	—	18.8	—	18.8	—
Desulfurized naphtha (bbl)	0.171	0.306	0.174	0.318	1.04	0.21	1.04	0.12
Kerosene (bbl)	0.260	0.532	0.260	0.532	—	—	—	—
Jet fuel (bbl)	—	—	—	—	—	0.807	—	0.921
Heavy gas oil (bbl)	0.382	—	0.392	—	—	—	—	—
Pitch (bbl)	0.204	0.214	0.214	0.220	—	—	—	—
(10^6 Btu/bbl)	6.55	6.55	6.33	6.33	—	—	—	—
Hydrogen consumed (lb)	4.50	6.10	5.30	6.9	8.0	6.4	12.3	10.7
Fresh heat (10^6 Btu)	0.112	0.152	0.128	0.162	0.179	0.143	0.276	0.240
Waste heat (10^6 Btu)	0.176	0.240	0.200	0.254	0.196	0.165	0.302	0.262
Steam (lb)	27.4	27.4	45.5	45.5	20	25	71	92
Condensate (gal)	2.16	2.94	2.46	3.12	2.4	3.0	8.5	11.0
Cost ($)	0.427	0.501	0.463	0.279	0.5893	0.4786	0.6093	0.4986

Sources: Yields from the H-oil process are based on information in A. R. Johnson, S. B. Alpert, and L. M. Lehman, "Processing Residual Fractions with H-Oil," Oil and Gas Journal, vol. 66, no. 26 (24 June 1968), pp. 86–94.

Yields for the hydrocracking process represent an averaging of several experiences reported in the literature, in particular: W. L. Nelson, "Investment Costs, Hydrocracking and Hydrogen Manufacture," ibid., vol. 66, no. 29 (15 July 1968), pp. 141–42; idem, "Total Hydrocracking Investment Is Large," ibid., vol. 66, no. 34 (19 August 1968), p. 93; D. H. Stormont, "Sohio's Second Hydrocracker Provides Exceptional Flexibility," ibid., vol. 65, no. 14 (3 April 1967), p. 160; and C. H. Watkins and W. L. Jacobs, "How to Get High-Quality Jet Fuels," ibid., vol. 67, no. 47 (24 November 1969), pp. 94–95.

Note: The maximum gasoline processes are applied to the light gas oils only; the maximum jet fuel processes to the heavy gas oils only. See table 1.

[a] Per barrel of reduced crude for the H-oil process; per barrel of feed (gas oil) for hydrocracking.

This is done in chapters 6 and 9. Third, the other products of the refinery (which can be thought of as by-products of the production of the five major products) are not constrained but are credited in the objective function at the prices shown in table 11.

TABLE 11.   Product Prices and Specifications: Input Costs and Qualities with Indications of Subsequent Variations

Description	Price or cost/unit ($)	Quality assumptions and constraints (% sulfur)	Variations explored in later chaps.
Inputs			
Crude oil			
East Texas crude (*bbl*)	3.75	0.4	—
Arabian mix (*bbl*)	3.02	1.44	Cost and availability
Isobutane (*lb*)	0.0168	—	—
Fresh heat purchased			
Very low sulfur (*10⁶ Btu*)	0.661	0.5	Cost
Low sulfur (*10⁶ Btu*)	0.593	1	Cost
High sulfur (*10⁶ Btu*)	0.477	2	Cost
Water for cooling (*10³ gal*)	0.015ᵃ	—	Cost
Desalter (*10³ gal*)	0.075ᵃ	—	Cost
Boiler (*10³ gal*)	0.15ᵃ	—	Cost
By-products			
Refinery gas (*lb*)ᵇ	0.022	—	—
Polymer sold as petrochemical feed (*bbl*)	5.00	—	—
Low-sulfur coke (*ton*)	8.00	2.0	—
High-sulfur coke (*ton*)	4.00	3.3	—
Butane (*lb*)ᶜ	0.0185	—	—
Sulfur (*long ton*)ᵈ	20	—	Price
Gasoline stocks for petrochemical feed (*bbl*)	4.41	—	—

Note: Inputs and by-products discussed in later chapters are (1) fresh heat purchased, chap. 4; (2) water for cooling, chap. 4; and (3) gasoline stocks for petrochemical feed, chap. 6.

ᵃ No specific quality limits were put on water withdrawals. However, the costs are assumed to reflect increasingly severe quality requirements (hence higher treatment costs or different sources) for desalter and boiler water.

ᵇ Letter to the author from R. W. Upchurch of Humble Oil Co., Supply Department.

ᶜ "Isobutane Prices vs. Regular Butane," *Oil and Gas Journal*, 1 April 1968, p. 83.

ᵈ See, for example, *Chemical Week*, 15 September 1971, p. 25, where the price of sulfur in the Mid-Atlantic states is said to be $31–$33 per long ton. Since, as will be observed, sulfur recovery is chosen in the model at the $20 price, it would, a fortiori, be chosen at the higher price. But there is evidence that the publicly quoted prices are considerably above the "real" price. In "Dark Cloud on Sulfur's Horizon," *Chemical Week*, 10 February 1971, pp. 25–34, John M. Winton quotes a Bureau of Mines estimate of $18/long ton for the average 1970 price FOB Gulf ports and calculates $20.20/long ton as the price implied by the 1970 Wholesale Price Index. The industry "hotly denied" those estimates, maintaining that the price was $10/long ton higher.

IV

REFINERY RESIDUALS:
MODIFICATION AND DISCHARGE

In the last chapter, the product side of the refinery was discussed and the processing units, choices of input mix, and alternative paths available to the refiner were described. At the same time it was indicated that residuals are inevitably generated in the operation of these process units, but that the quantity and type generated will vary with process choices made in response to such influences as market demand for products and availability of inputs. In this chapter the alternatives available for dealing with these residuals once they have been generated are described; as background for this description of the details of the residuals-handling system of the refinery, residuals modification processes are first discussed more generally.

Once residuals have been generated the laws of conservation of mass and energy require that their mass and energy content be accounted for completely, if not by discharge to the natural world then by recycling or by-product production activities that make of them production inputs or outputs respectively.[1] It should be noted, of course, that while recycling and by-product production can be used to reduce the total discharges resulting from the original production process, their operation will also

---

[1] Recall that in chap. 2 residuals generation is defined as being measured subsequent to any recycling and by-product recovery operations that would be carried out at existing prices (for inputs, products, etc.) *in the absence of* actions taken by public authorities to control discharges. Thus in the catalytic cracker catalyst regenerator, particulate residuals generation is measured after the recovery unit that is installed as standard practice because the catalyst is sufficiently expensive to warrant the recovery of fines from the combustion gas stream.

imply the generation of residuals referred to in chapter 2 as secondary residuals. These arise because of the necessity of adding energy (most often heat) or chemicals in order to obtain the desired reactions. It is, then, by no means certain that the ultimate mass and energy content of the plant's discharges will be reduced by these processes. In their absence, however, it *is* certain that the total residual mass and energy to be discharged must be at least as great as that generated, though the pattern of the discharges by type, location, timing, and natural media used may be greatly altered from that implied by direct discharge from the production process.

Specifically, the four basic varieties of modification processes are:

1. One that changes the form or type of residual (chemical compound or physical state). A very simple example of such a modification is the flaring (open burning) of hydrogen sulfide. This process converts the poisonous and foul-smelling hydrogen sulfide to sulfur oxide plus water vapor before discharge. Notice also that it involves an increase in the weight of the residual discharged, because of the addition of oxygen in burning, and an additional heat residual, because of the exothermic oxidation reaction. This trade-off is at the heart of many of the processes that transform residual types. The discharge of a more noxious and poisonous substance is avoided by settling for a greater weight of a somewhat less noxious and poisonous one, plus some quantity of other residuals. Other examples of type transformations are (a) the burning of municipal solid waste where the volume and weight of solids requiring landfill disposal are reduced at the cost of increased total residuals weight and total energy released;[2] and (b) the neutralization of bases or acids to produce a neutral solution and salts, either dissolved or precipitated.

2. A second possible modification is in the natural media to which the residual will be discharged.[3] Thus, for example, solids suspended in process water may be discharged to a water course or settled out and subsequently "discharged" onto land. Hydrogen sulfide dissolved in water may be discharged or scrubbed out and discharged to the air as a gas.

[2] Trash burning also involves a change of the receiving medium, the second type of modification process discussed here.
[3] It is implicitly assumed that in the absence of any modification, a liquid residual or a residual dissolved or suspended in a liquid would be discharged to a body of water; a gaseous residual or a residual suspended in a gas stream, discharged to the atmosphere; and a solid residual, discharged to the land. This assumption is accurate enough for expository purposes, but for application to specific residuals from specific plants, it must be examined.

Notice in this connection that many recirculation processes involve recycling of residuals transport media, not of residuals themselves. In these processes the residuals are generally removed to some degree from the stream to be recirculated and then discharged into a different medium. Thus cooling towers do not decrease the total quantity of residual heat from a process; rather they remove it from cooling *water* and release it directly to the air through the evaporation of some part of that water. As a general rule the word "recirculation" is used here for processes involving residual transport streams (usually with a change in the residual's discharge medium and sometimes in the residual's form as well); the word "recycling" is restricted to processes in which residuals themselves are returned to the input side of the production processes. Some processes will combine features of each of these types.

3. A third possibility is to discharge the same residual to the same medium but at a different location. For example, liquid residuals might be piped out to sea instead of being discharged into a river or estuary; gases and particulates may be vented from tall stacks rather than short ones (so that winds will tend to disperse dangerous or obnoxious concentrations before they drift down to ground level).

4. A final general modification alternative is to arrange for the discharge of the same residual to the same medium at the same location but in a different time pattern. Thus, for example, liquid residuals may be stored in lagoons or tanks during periods of low streamflow and discharged during periods of high flows.[4]

It is worth stressing once more that modification processes make it possible to alter several important dimensions of residuals discharges, but they do not make residuals disappear. For any real problem, however, some residuals—either originally generated or produced in some modification process—may be ignored; strict material and energy balances will not be maintained and it will *seem* that residuals have conveniently disappeared. Which residuals may be ignored will depend on the particular problem under study and especially on its time and space dimensions, as was stressed in chapter 2. Thus the fact that carbon dioxide (CO_2) emissions to the atmosphere are ignored in regional studies with limited time

---

[4] The corresponding techniques for gaseous residuals generally involve fuel (input) substitution or changes in production schedules and thus are, in my lexicon, changes in generation rather than modification. Generally, timing modifications must involve storage of residuals already generated and the practicality of the step will be dependent on storage volume necessary.

horizons must not be allowed to lull analysts into forgetting that they exist and are quantitatively very large for any process involving the burning of fossil fuel.

In the refinery example most of the modification processes are of the first two varieties: residual form, or type, and discharge medium, but a by-product production alternative is also included. A fairly limited set of residuals is followed and, in particular, CO_2, carbon monoxide (CO), water vapor, and energy discharges to the atmosphere are ignored. As a reminder of the sources of and influences on residuals generation in the refinery, figure 11 shows each process unit, the kinds of residuals generated in it, and the major choices available *in the model* that affect the quantities of residuals generated per barrel of input or output.

WATERBORNE RESIDUALS

The refinery produces three major types of residual-bearing water streams: cooling water that carries residual heat away from the process units; spent caustics and acids from the gasoline-sweetening unit and the alkylation unit respectively; and process steam condensate from various process units containing dissolved sulfide, phenols, and entrained oil with some biological oxygen demand.[5] Each of these streams is discussed separately, but included in the residuals-modification system is a recirculation alternative linking process water and the cooling system. The overall residuals-handling system is shown schematically in figure 12.

Cooling Water

In chapter 2 the generation of residual heat per barrel of input (or output) was estimated for each process unit. The influence on this generation of various choices, such as capital intensity, the recycle ratio in the cat crackers, and the operating severity in the reformer was also considered. Once generated, the residual heat is assumed to be carried away by cool-

---

[5] The model is limited to sulfide, ammonia phenols, oil, and BOD, but actual refinery residuals would include suspended and dissolved solids, and certain heavy metals. Three considerations influenced this choice: the desire to include residuals, such as sulfide, with strong linkages to other discharge media; limitations on the ability of existing environmental models to handle exotic residuals such as heavy metals; and a need to simplify at some point in order to make the task of model construction a finite one. "BOD" refers to biological oxygen demand measured over five days—sometimes written BOD_5. (Note that phenols generally contribute to BOD.)

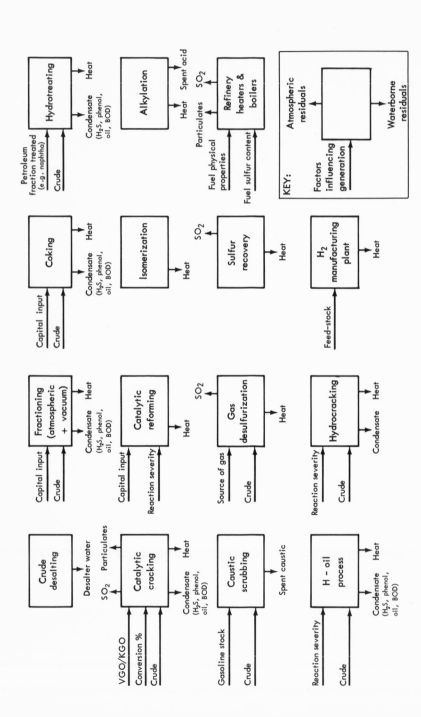

Figure 11. Process residuals and factors under the control of the refiner.

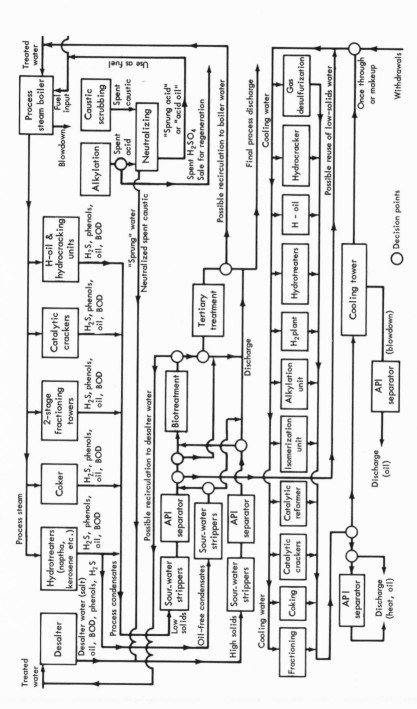

Figure 12. Flow diagram of liquid residuals streams and possible modification processes.

ing water that is allowed to rise 35°F in temperature in the coolers and hence is capable of removing 35 Btu/lb, or 0.292 10^6 Btu/1,000 gallons (gal). In the simplest case, when there is no pressure on the refinery to reduce either its water withdrawals or its residual heat discharges, this cooling water is supplied on a once-through basis at an assumed cost of $0.015/1,000 gal. (This cost is low to reflect the fact that once-through cooling water may be of low quality and hence requires very little or no pretreatment even when withdrawn from polluted sources.)

It is assumed that cooling water used once-through picks up from miscellaneous leaks an average of 5.3 parts per million (ppm) of oil in the course of performing its cooling duty. This figure is based on data in a survey of domestic refinery effluents conducted by the American Petroleum Institute (API) in 1967.[6] For the benchmark case (see chapter 5), it is assumed that this cooling water may be discharged without any reduction in its oil content. To allow for response to constraints or charges on oil discharges, however, the model provides the alternative of building an API oil-water separator in which 80 percent of the oil can be removed.[7] The recovered oil-water mixture is assumed to be burned within the refinery and to provide 10,000 Btu/lb useful heat. The separator is costed at $0.064/1,000 gal on the basis of a capacity of 85 million gal/day. This size figure represents a rough estimate of the volume of dirty cooling water for a refinery having a capacity of 150,000 bbl/day (i.e., that water coming from process units in which the heavier oil fractions are involved). In particular, cooling water from the coker and cat cracker are assumed to have by far the greatest part of the residual oil, and the cost of the API unit is assumed to reflect the cost of segregating these streams as well as of treating them.[8]

[6] API, Committee for Air and Water Conservation, *1967 Domestic Refinery Effluent Profile* (September 1968), table Ib and IIb. From the data on the quantities of oil and volumes of water discharged, it was estimated that average cooling water discharges for the refinery type considered here were about 1,050 gal/bbl of crude charged, with an average oil content of 5.3 ppm. Process water discharges averaged 25.1 gal/bbl with 76 ppm of oil.
[7] An API separator is simply a long concrete basin, or set of basins, through which the waste water flows at a rate calculated to allow some chosen percentage of oil droplets to rise to the surface. At the outlet end, skimmers remove the floating oil. This removal percentage is based on M. R. Beychok, *Aqueous Wastes from Petroleum and Petrochemical Plants* (John Wiley, 1967), pp. 225–33; and the Bechtel Corporation's technical review of an earlier draft of this study.
[8] See Beychok, *Aqueous Wastes*, p. 29; and API, *Manual on Disposal of Refinery Wastes: Volume on Liquid Wastes* (New York, 1969), fig. 3–1. The lighter hydrocarbons are assumed to evaporate from the cooling water to the extent that they leak into it at all.

If there is pressure to reduce either withdrawals or heat discharge, the refiner is free to choose to install cooling towers. The cooling system then becomes semiclosed with makeup withdrawals estimated to be 9.25 percent of gross cooling flow.[9] This makeup replaces water lost through evaporation (the method of cooling the circulating water) amounting to 35 gal/1,000 gal recirculated, or 3.5 percent.[10] Windage (water not evaporated but blown out of the towers as mist) is assumed to be 0.75 percent, and blowdown (necessary to control the concentration of solids and oil in the circulating water), 5 percent. Makeup may be provided either from the original source of cooling water, or, as discussed below, from certain process water streams after initial treatment to remove sulfide and oil. The total costs of cooling tower use are $0.0356/1,000 gal, exclusive of the cost of makeup water.

When recirculation of primary treatment effluent to cooling tower makeup is practiced (see below), the makeup feed must be treated to inhibit the growth of algae in the cooling system. These costs are not included in the overall cooling tower costs, but are attached to the recirculation vector and are assumed to be $0.04/1,000 gal.

The cooling tower blowdown contains relatively high concentrations of oil, dissolved solids, and residual chemicals from makeup pretreatment. These high concentrations result from the evaporation of water—but not of solids, chemicals, or oil—in the tower. I do not deal with the solids or chemicals in the model, even though they can pose significant residuals management problems.[11] This decision was made primarily on the grounds of simplifying the technical research task, since the serious residuals problems with the blowdown involve the pretreatment chemicals that include algicides and that require complex predischarge modification to alter the type of compound discharged. It is assumed that the dissolved solids are not generally a serious problem in themselves, though this may be incorrect at certain locations. The oil residual, however, is followed. The basic assumption is that no oil is evaporated in the tower.

[9] My estimates of the costs and operating characteristics of cooling towers are based on information contained in Paul H. Cootner and George O. G. Löf, *Water Demand for Steam Electric Generation* (RFF, 1965), pp. 60–75; and Ian Wigham, "Designing Optimum Cooling Systems," *Chemical Engineering*, 9 August 1971, pp. 95–102.

[10] In evaporating, 1 lb of water removes 1,000 Btu from the circulating stream. Hence 292 lb, or 35 gal, must be evaporated to cool 1,000 gal through 35 degrees.

[11] See "Treating Blowdown from Cooling Towers," *Industrial Water Engineering*, vol. 6, no. 6 (June 1969), pp. 33–35, for a discussion of blowdown treatment problems. For a more optimistic view, compare John M. Donohue, "Making Cooling Water Safe for Steel and Fish, Too," *Chemical Engineering*, 4 October 1971, pp. 98–102.

Hence, the oil in the blowdown must equal the oil in the discharge of once-through cooling water; or the concentration of oil in the blowdown must be 20 times that in once-through discharge—about 106 ppm. Because the oil concentration of this stream is sufficiently high, it is assumed that the blowdown must go to an API separator where, again, 80 percent of the oil is removed. Here the cost is based on 5 percent of the total cooling flow or about 8.5×10^6 gal/day.[12] This gives a total average cost of $0.122/1,000 gal treated.

Spent Caustics and Acids

The model is constructed to require the caustic scrubbing of gasolines from the catalytic cracker and to allow but not require it for straight-run gasoline. In the process of "sweetening" these gasolines by removing sulfur, organic sulfur compounds (mercaptans), and complex aromatic hydrocarbons, the caustic (NaOH) solution is spent, and the resulting weak caustic contains as residuals the impurities removed from the gasoline. At the same time, operation of the alkylation unit produces a stream of spent sulfuric acid catalyst. Since the first step in handling the caustic stream must be neutralization, the model provides for the use of the spent acid as the neutralizing agent.[13] Spent acid not required for this task is assumed salable to a reprocessor for $0.004/lb.[14] The caustic neutralization produces two streams that separate rather easily because of density differences. One is called "sprung acid" and consists of essentially all the oil and mercaptans and most of the phenols from the spent caustic along with some of the hydrogen sulfide, water, and sodium sulfate (the salt produced in neutralization). The other stream, "sprung water," is primarily water (it contains most of the water from the two spent aqueous solutions) plus a small quantity of phenols, some H_2S, and a large amount of sodium sulfate.

The calculations necessary to go from the known characteristics of a raw gasoline and the desired impurity removal to the composition of the spent caustic, and hence of the sprung acid and water streams, are tedious

[12] The assumption is that the cooling water recirculation system will not be segregated; thus all the blowdown will have to be treated.

[13] An alternative method of dealing with spent caustics is to sell them. Several firms in the United States now buy refinery-spent caustics and recover phenols (including phenol, cresol, and xylenol) from them. See "Waste Reprocessing: Doomed by Lead-free Gas," *Chemical Week*, vol. 107, no. 26 (23 December 1970), pp. 33–34.

[14] Based on W. L. Nelson, *Guide to Refinery Operating Costs* (Tulsa, Okla.: Petroleum Publications, 1970), p. 89, where spent acid is credited at $6–$10 per ton.

but straightforward, and their exposition here would serve no particular purpose.[15] Beychok, however, suggests two rules of thumb that help considerably in understanding the net effect of this operation on the refinery's waterborne residuals situation. He observes that the phenol content of sprung water can be assumed to be 0.8 percent (8,000 ppm) by weight, and the H_2S content, 0.3 percent (3,000 ppm), independent of the gasolines scrubbed.[16] I have taken advantage of these approximate rules to assume that the sprung water from the several scrubbing alternatives may be treated as a single stream of homogeneous composition.[17] This stream is then handled very much as are the foul condensates from the process units, with certain exceptions mentioned below.

The sprung acid is assumed to be burned in a refinery boiler designed to take sludges, oil emulsions, and so forth. Its heat content is estimated to be 8,800 Btu/lb, and allowance is made for its high average sulfide and mercaptan content by assuming that 0.448 lb of SO_2 are emitted per pound of sprung acid burned.[18]

Process Condensates

Process condensates are those water streams that result from the condensation of steam that has been in direct contact with petroleum stocks as part of normal refinery processing. Thus, for example, steam introduced into a fractioning tower to aid in more exact separation of desired boiling ranges condenses with the overhead gases from the tower and forms part of the condensate of that process.[19] Given this origin, it is not surprising that process condensates generally constitute a serious residuals problem for the refinery, having picked up along with some of the oil itself, hydrogen sulfide, phenols, and other unstable organics (oxygen demanders). These are the principal residuals found in the condensates, except for ammonia, which is usually injected into the condensing system to control the acidity caused by dissolved hydrogen sulfide as well as being formed from nitrogen in the air and in the oil in hydrotreating and cracking units.

[15] See Beychok, *Aqueous Wastes*, pp. 58–74, especially pp. 72–74.
[16] Ibid., pp. 68 and 74.
[17] This is possible because I ignore the sodium sulfate salt.
[18] The average hydrogen sulfide content of the sprung acid is assumed to be 23.8 percent. Hence the sulfur content is 22.4 percent, and the SO_2 produced by oxidizing this sulfur amounts to twice the sulfur content per pound.
[19] The uses of steam in refining were discussed in chap. 3.

In the model the condensate volumes are based on Nelson's estimates of process steam required in each process unit under the assumption that 100 percent of this steam appears as condensate.[20] For phenols, oil, and BOD, concentrations have been chosen on the basis of data reported in Beychok for condensates from similar process units so that my figures can be said to be roughly reasonable in that they are within the range of observed experience (and there appears to be substantial variation in the real world).[21] But they are not connected in any rigorous way with the characteristics of the two alternative crudes nor with the specific operating conditions implicitly assumed for each unit. For hydrogen sulfide, on the other hand, an effort has been made to calculate condensate concentrations fairly carefully on the basis of the sulfur content of the process-unit feedstock, the fate of that sulfur in the process, and the solubility of H_2S in the sour condensate.[22] Ammonia is assumed to amount to 75 percent of the hydrogen sulfide in the raw condensates. The resulting figures, then, are related explicitly to feedstock and conditions in the process unit. The calculated and assumed residuals concentrations for the various condensate streams are shown in table 12 under "Raw loads generated in process units." A larger-scale study of refinery residuals would involve such calculations for phenols, oil, and ammonia, but the approach would be the same.

In the model, the process condensates plus sprung water and desalter effluent may be handled in a variety of ways, depending on the residuals management pressures impinging on the refiner. The following description of these alternatives applies to nearly all the individual streams identified in table 12, but exceptions are noted. The streams subject to common modification alternatives might join a common refinery condensate sewer and be treated in the same large-scale units. For the streams that are treated differently, separate sewers or treatment facilities (or both) are necessary and are reflected in the costs assumed. Note that in order to reflect the distinct compositions of the various process-unit condensates and the effects on these of the different modification alternatives, each stream is followed separately (using volume as the operational variable) up through the biological treatment alternative.

[20] Nelson, *Guide to Operating Costs*. The specific processes were described in chap. 3 of the present study. As noted there, condensates from the reformer, isomerization unit, and alkylation unit are assumed to be free of residuals and are discharged directly. This assumption is based on data provided in Beychok, *Aqueous Wastes*.

[21] Beychok, *Aqueous Wastes*.

[22] These calculations rely heavily on the material in ibid., pp. 48–58.

The general method was described in chapter 2, as was the problem of the choice of a cost figure when the scale is unknown.[23]

As a first step the condensates are sent to the sour-water stripper in which steam is used to remove 99 percent of the dissolved H_2S and 90 percent of the ammonia (NH_3).[24] The rate of steam use is 2 lb/gal of condensate. The resulting H_2S gas, along with any from the gas desulfurization plant described in chapter 3, may either be flared—that is burned in the atmosphere to produce an SO_2 emission—or sent to a sulfur recovery unit. If the former route is chosen, each pound of H_2S produces 1.88 lb of SO_2 emission to the atmosphere. In the recovery plant, on the other hand, 90 percent of the sulfur in the H_2S is recovered, and only 10 percent is lost (presumably as SO_2) to the atmosphere. Thus SO_2 emissions resulting from gas desulfurization and the removal of H_2S from water may be cut to 0.188 lb/lb of H_2S through sulfur recovery.

The complete sulfur recovery process vector is:[25]

H_2S in (1,000 scf)	89.5 lb
Sulfur out	75.6 lb[26]
Usable heat produced	0.183×10^6 Btu
Steam required	184 lb
Residual heat	0.035×10^6 Btu
SO_2 emitted	16.8 lb
Cost	\$0.355

After stripping, the condensates are passed through an API oil-water separator that is assumed to remove 80 percent of the oil. The recovered oil is assumed to be burned in the refinery sludge furnace and to yield 10,000 Btu/lb net heat.

The desalter water is assumed to bypass the sour-water stripper but to require an individual separator because of its relatively high solids load.

[23] This technique, of course, involves one in the problem of scale economies already discussed. It also results in the model choosing different levels of treatment for different streams under particular charges or discharge limits. This is not consistent with the common-sewer, common-treatment process assumption.

[24] Ibid., pp. 175–81. Sour-water stripping is required even in the absence of formal residuals management actions; the refiner is assumed to feel he cannot release to a watercourse any significant amount of H_2S because of its odor and toxicity. This assumption appears to jibe with the observed sulfide discharges per barrel of crude reported in U.S. Department of the Interior, FWPCA, *The Cost of Clean Water*, vol. 3, *Industrial Waste Profile No. 5*, "Petroleum Refining" (November 1967), table 6, app. A; and API, *1967 Effluent Profile*, table Ib. See also table 20 in the present study.

[25] Process utilities, residuals, cost, etc., are based on W. L. Nelson, "Cost of Sulfur Manufacture," *Oil and Gas Journal*, 2 September 1968, pp. 101–02.

[26] Notice that 89.5 lb of H_2S contain 84 lb of sulfur. Hence 90 percent recovery produces 75.6 lb of sulfur.

TABLE 12. Residiuals Characteristics of Process Condensate

(pounds per 1,000 gallons)

Streams[a]	Raw loads generated in process units					Residuals loads after required primary treatment[b]				
	H₂S	NH₃	Phenol	Oil	BOD	H₂S	NH₃	Phenol	Oil	BOD
Desalter water, 2.5 gal/bbl	0.12	0.09	0.12	1.88	2.34	neg.	0.01	0.12	0.38	2.34
ETC fractioning, 5.4 gal/bbl	1.02	0.76	0.83	0	1.42	0.01	0.08	0.83	0	1.42
AC fractioning, 5.4 gal/bbl	36.70	27.50	1.25	0	2.08	0.37	2.75	1.25	0	2.08
Coker ETC, 15 gal/bbl	35.53	26.60	3.34	1.25	5.59	0.36	2.66	3.34	0.25	5.59
Coker AC, 15 gal/bbl	112.60	84.50	4.17	1.25	7.00	1.13	8.45	4.17	0.25	7.00
ETC cat cracker, VGO, L.C., 4.2 gal/bbl	22.68	17.00	4.59	0.42	7.67	0.23	1.70	4.59	0.08	7.67
ETC cat cracker, VGO, H.C., 9 gal/bbl	14.84	11.13	4.59	0.42	7.67	0.15	1.11	4.59	0.08	7.67
AC cat cracker, VGO, L.C., 4.2 gal/bbl	77.31	58.00	4.59	0.42	7.67	0.77	5.80	4.59	0.08	7.67
AC cat cracker, VGO, H.C., 9 gal/bbl	51.29	38.50	4.59	0.42	7.67	0.51	3.85	4.59	0.08	7.67
ETC cat cracker, KGO, L.C., 4.2 gal/bbl	72.60	54.50	8.34	0.83	13.93	0.73	5.45	8.34	0.17	13.93
ETC cat cracker, KGO, H.C., 9 gal/bbl	51.00	38.20	8.34	0.83	13.93	0.51	3.82	8.34	0.17	13.93
AC cat cracker, KGO, L.C., 4.2 gal/bbl	226.00	169.50	8.34	0.83	13.93	2.26	16.95	8.34	0.17	13.93
AC cat cracker, KGO, H.C., 9 gal/bbl	150.0	120.0	8.34	0.83	13.93	1.50	12.00	8.34	0.1	13.93
Sprung water[d]	25.02	0	66.72	0	111.76	0.25	0	66.72	0	111.76
ETC naphtha hydrotreater, 0.66 gal/bbl	96.30	72.10	0	0	1.00	0.96	7.21	0	0	1.00
AC naphtha hydrotreater, 1.32 gal/bbl	264.0	198.0	0	0	1.00	2.64	19.80	0	0	1.00
ETC kerosene hydrotreater, 1.20 gal/bbl	154.0	115.5	0	0.1	10.00	1.54	11.55	0	0.02	10.00

AC kerosene hydrotreater, 2.40 gal/bbl	289.0	226.0	0	0.1	10.00	2.89	22.60	0	0.02	10.00
ETC coker gasoline hydrotreater, 0.84 gal/bbl	174.0	130.5	0	0.3	30.00	1.74	13.05	0	0.06	30.00
AC coker gasoline hydrotreater, 1.68 gal/bbl	239.0	179.0	0	0.3	30.00	2.39	17.90	0	0.06	30.00
AC high-S, dist. F.O. hydrotreater, 4.20 gal/bbl	705.0	529.0	0	0.6	50.00	7.05	52.90	0	0.12	50.00
ETC resid. 1, hydrotreater, 3 gal/bbl	501.0	376.0	0	1.0	80.00	5.01	37.60	0	0.20	80.00
ETC resid. 2, hydrotreater, 3 gal/bbl[e]	575.0	430.0	0	1.0	80.00	5.75	43.00	0	0.20	80.00
AC resid. 3, hydrotreater, 5 gal/bbl	865.0	649.0	0	1.0	80.00	8.65	64.90	0	0.20	80.00
AC resid. 4, hydrotreater, 5 gal/bbl	920.0	690.0	0	1.0	80.00	9.20	69.00	0	0.20	80.00
ETC H-oil, L.C., 2.16 gal/bbl	334.0	250.0	1.00	0.40	4.00	3.34	25.00	1.00	0.08	4.00
ETC H-oil, H.C., 2.94 gal/bbl	334.0	250.0	1.00	0.20	2.00	3.34	25.00	1.00	0.04	2.00
AC H-oil, L.C., 2.46 gal/bbl	334.0	250.0	1.00	0.40	4.00	3.34	25.00	1.00	0.08	4.00
AC H-oil, H.C., 3.12 gal/bbl	334.0	250.0	1.00	0.20	2.00	3.34	25.00	1.00	0.04	2.00
ETC hydrocracker max. gasoline, 2.40 gal/bbl	258.0	194.0	0.50	0.10	1.00	2.58	19.40	0.50	0.02	1.00
AC hydrocracker max. gasoline, 8.5 gal/bbl	294.0	220.0	0.50	0.10	1.00	2.94	22.00	0.50	0.02	1.00
ETC hydrocracker max. gasoline, 3 gal/bbl	276.0	207.0	1.00	0.20	2.00	2.76	20.70	1.00	0.04	2.00
AC hydrocracker max. jet fuel, 11 gal/bbl	334.0	250.0	1.00	0.20	2.00	3.34	25.00	1.00	0.04	2.00

See notes at end of table.

TABLE 12. (Continued)

(pounds per 1,000 gallons)

Streams[a]	Residuals loads after use of primary effluent as cooling tower makeup			Residuals after activated sludge treatment			
	Oil	BOD	NH$_3$	Phenol	NH$_3$	Oil	BOD
Desalter water, 2.5 gal/bbl	[c]	[c]	[c]	0	0	0.15	0.52
ETC fractioning, 5.4 gal/bbl	0	0.57	0.05	0.03	0.04	0	0.31
AC fractioning, 5.4 gal/bbl	0	0.83	1.65	0.05	1.38	0	0.46
Coker ETC, 15 gal/bbl	0.25	2.24	1.60	0.13	1.33	0.10	1.23
Coker AC, 15 gal/bbl	0.25	2.80	5.06	0.17	4.22	0.10	1.54
ETC cat cracker, VGO, L.C., 4.2 gal/bbl	0.08	3.07	1.02	0.18	0.85	0.03	1.69
ETC cat cracker, VGO, H.C., 9 gal/bbl	0.08	3.07	0.66	0.18	0.56	0.03	1.69
AC cat cracker, VGO, L.C., 4.2 gal/bbl	0.08	3.07	3.48	0.18	2.90	0.03	1.69
AC cat cracker, VGO, H.C., 9 gal/bbl	0.08	3.07	2.31	0.18	1.92	0.03	1.69
ETC cat cracker, KGO, L.C., 4.2 gal/bbl	0.17	5.55	3.27	0.33	2.72	0.07	3.06
ETC cat cracker, KGO, H.C., 9 gal/bbl	0.17	5.55	2.29	0.33	1.91	0.07	3.06
AC cat cracker, KGO, L.C., 4.2 gal/bbl	0.17	5.55	10.20	0.33	8.48	0.07	3.06
AC cat cracker, KGO, H.C., 9 gal/bbl	0.17	5.55	7.2	0.33	6.00	0.07	3.06
Sprung water[d]	[c]	[c]	[c]	2.67	0	0	24.60
ETC naphtha hydrotreater, 0.66 gal/bbl	0	0.40	4.33	0	3.60	0	0.22
AC naphtha hydrotreater, 1.32 gal/bbl	0	0.40	11.90	0	9.90	0	0.22
ETC kerosene hydrotreater, 1.20 gal/bbl	0.02	4.00	6.94	0	5.78	0.01	2.20
AC kerosene hydrotreater, 2.40 gal/bbl	0.02	4.00	13.60	0	11.30	0.01	2.20
ETC coker gasoline hydrotreater, 0.84 gal/bbl	0.06	12.00	7.84	0	6.52	0.02	6.60
AC coker gasoline hydrotreater, 1.68 gal/bbl	0.06	12.00	10.70	0	8.95	0.02	6.60
AC high-S, dist. F.O. hydrotreater, 4.20 gal/bbl	[c]	[c]	[c]	0	26.40	0.05	11.00
ETC resid. 1, hydrotreater, 3 gal/bbl	[c]	[c]	[c]	0	18.80	0.08	17.60
ETC resid. 2, hydrotreater, 3 gal/bbl[e]	[c]	[c]	[c]	0	21.50	0.08	17.60
AC resid. 3, hydrotreater, 5 gal/bbl	[c]	[c]	[c]	0	32.45	0.08	17.60
AC resid. 4, hydrotreater, 5 gal/bbl	[c]	[c]	[c]	0	34.50	0.08	17.60
ETC H-oil, L.C., 2.16 gal/bbl	0.08	1.60	15.00	0.04	12.50	0.03	0.88
ETC H-oil, H.C., 2.94 gal/bbl	0.04	0.80	15.00	0.04	12.50	0.02	0.44

AC H-oil, L.C., 2.46 gal/bbl	0.08	1.60	15.00	0.04	12.50	0.03	0.88
AC H-oil, H.C., 3.12 gal/bbl	0.04	0.80	15.00	0.04	12.50	0.02	0.44
ETC hydrocracker max. gasoline, 2.40 gal/bbl	0.02	0.40	11.60	0.02	9.70	0.01	0.22
AC hydrocracker max. gasoline, 8.5 gal/bbl	0.02	0.40	13.20	0.02	11.00	0.01	0.22
ETC hydrocracker max. gasoline, 3 gal/bbl	0.04	0.80	12.40	0.04	10.35	0.02	0.44
AC hydrocracker max. jet fuel, 11 gal/bbl	0.04	0.80	15.00	0.04	12.50	0.02	0.44

Note: ETC = East Texas crude; AC = Arabian mix crude; VGO = virgin gas oil; KGO = coker gas oil; H.C. = high-conversion rate; and L.C. = low-conversion rate.

a Volume given in gallons per barrel of feed to the unit in question. Thus the naphtha hydrotreater, when operating on the low-sulfur naphtha, generates 0.66 gal of condensate per barrel of naphtha.

b These loads are calculated for the original volume of water. In fact the volume has been increased for the stripped condensates by a factor of 1.24. Thus the actual concentrations would be lower. The numbers are presented this way to facilitate comparisons. A similar comment holds for the later discharge stages.

c The stream is too high in solids to be used as cooling tower makeup.

d Sprung water volume varies with the gasoline scrubbed.

e This loading is also assumed to characterize the condensate from the hydrotreating of medium-sulfur distillate fuel oil.

The sprung water is also assumed to be kept separate through the primary treatment stages because of a heavy solids load, but it requires stripping only, having no oil content. Hence the unit costs for each of these streams is higher than that for the condensates.

Costs for these units are summarized here, along with the sizing assumptions on which the costs are based:

1. General refinery sour-water stripper installation: 2.5×10^6 gal/day; total average cost, $0.160/1,000 gal.
2. API separator for oily, sour condensates: 3.1×10^6 gal/day (accounting for the addition of 0.24 gal of water, as steam, per gallon stripped); total average cost, $0.162/1,000 gal.
3. Sour-water stripper for desalter water: 0.375×10^6 gal/day; average total cost, $0.275/1,000 gal.
4. API unit for desalter water: 0.465×10^6 gal/day; average total cost, $0.28/1,000 gal.
5. Sprung water is assumed to be stripped in the unit serving desalter water at the same average total cost.

After passing through the API separator the condensate may go in any one of three directions: it may be discharged; it may be used as makeup feed to the cooling tower, if one is in operation;[27] or it may be sent on to a secondary biological treatment process, assumed here to be the activated sludge process. If the condensate is discharged, the remaining concentrations of sulfide ammonia, phenol, oil, and BOD are used by the program to calculate the weight of residuals discharged. If the condensate goes as makeup to the cooling towers, similar provision is made for reflecting residuals in the tower blowdown. For this purpose it is assumed that all the remaining sulfide and phenol, 40 percent of the ammonia, and 60 percent of the BOD, but none of the oil, are "removed" in the process of circulation through the cooling system.[28]

If the condensate is sent for further modification, it enters an activated sludge process that is assumed to remove all the remaining sulfide, 96

[27] Desalter and sprung water may not be sent to the cooling system.
[28] Based on Beychok, *Aqueous Wastes*, pp. 264–65. Note that here, for simplicity, the subsequent forms these residuals take are ignored. Thus phenols may be oxidized in the towers, but hydrocarbons may remain in the water or may be evaporated to the atmosphere. The sulfide and the rest of the oxygen-demanding organics are handled similarly. Since, however, the total quantity of sulfide that could be oxidized in towers is less than about 400 lb/day, which would produce about 800 lb/day of SO_2, the distortion involved is small. The total SO_2 emission from the refinery chosen for the benchmark case is over 214,000 lb/day.

percent of the phenols, 60 percent of the remaining oil, 50 percent of the ammonia, and 78 percent of the BOD. The cost of this process is estimated to be \$0.236/1,000 gal.[29] In the process a sludge is produced with a dry solid content equal to 0.75 lb/lb of BOD removed (or 0.585 lb/lb of original BOD).[30] This sludge will ordinarily be subject to further modification before the ultimate residuals are discharged. Digestion followed by drying and landfill or thickening followed by dewatering and incineration are perhaps the two major possible routes. For simplicity, only the latter alternative is included in the model. The basic assumptions made in constructing the sludge-handling vectors are:[31]

1. The original sludge is 5 percent solids, and the cake formed by thickening and dewatering is 20 percent solids. The cost of dewatering is \$0.15/100 lb of cake.
2. The solids are half volatile (combusted or driven off in incineration) and half inert. Of the inert solids, 62.4 percent show up as furnace ash in the incinerator, while 37.6 percent come off as particulates in the flue gases. The cost of incineration is \$0.25/100 lb of cake burned, in the absence of particulate controls.
3. For an extra \$0.038/100 lb, particulate emissions may be reduced by 80 percent; and for \$0.075/100 lb (over the no-control cost) they may be reduced by 95 percent.

After biological treatment, another choice among three routes is available. The plant effluent may be recirculated to the desalter at a cost of \$0.04/1,000 gal; it may be discharged; or it may be sent to yet another treatment stage—activated carbon absorption. If discharge is chosen, total quantities of residuals are determined as described above from

[29] Based on material on water use in petroleum refining prepared for RFF by Herbert Mohring and J. Hayden Boyd; Robert Smith, "Cost of Conventional and Advanced Treatment of Waste Water," *Journal of the Water Pollution Control Federation*, vol. 40, no. 9 (September 1968), pp. 1546–74; API, *Disposal of Refinery Wastes*, chap. 13; and Beychok, *Aqueous Wastes*, pp. 257–63, 275–80.

[30] Note that 1 lb of BOD is a quantity of organic material consuming 1 lb of oxygen over some set time span at a set temperature in oxidizing to a more stable chemical form. The original weight of organics need not be equal to a pound. If, for example, the organic material consisted of pure carbon, the weight consuming 1 lb of oxygen in complete oxidation would be 0.375 lb.

[31] Based on information in P. J. Cardinal, Jr., "The Incinerator's Role in Sludge Disposal," *The American City*, December 1967, pp. 108–10; G. T. Hopkins and R. L. Jackson, "Solids Handling and Disposal," *Public Works*, January 1968, pp. 607–28; and R. S. Burd, "Final Report: A Study of Sludge Handling and Disposal" (submitted to FWPCA, June 1966, mimeo.). Without particulate removal, costs of sludge handling are then \$0.020 per pound of dry solids, or \$40/ton.

volumes and concentrations. If tertiary treatment is chosen, a heat input of 7,600 Btu/1,000 gal of condensate is required in the carbon regeneration furnace, and the cost per 1,000 gal is estimated at $0.35.[32] I assume for convenience that the average tertiary effluent will contain 0.004 lb of phenols, 0.083 lb of BOD, 0.25 lb of ammonia, and 0.05 lb of oil per 1,000 gallons, or 0.48 ppm, 10 ppm, 40 ppm, and 6 ppm respectively.

The effluent from the activated carbon absorption unit is assumed to be clean enough for recirculation to the refinery steam boilers after a cost of $0.08/1,000 gal for additional chemicals.[33]

AIRBORNE RESIDUALS

Although the refinery has a significant problem with emissions of gases and particulates to the atmosphere, the nature and distribution of the sources of these residuals and the choices made in constructing the model result in a considerably less complex system than that just described for waterborne residuals. One of the major potential sources of refinery emissions of sulfur oxides has already been described—the flaring of H_2S stripped from refinery gases and sour condensates. The SO_2 and particulate emissions from the cat cracker and the particulates generated by sludge incineration have also been mentioned. Figure 13 shows these sources and the others described below, along with the modification options available.

Cat-Cracking Catalyst Regenerators

As described in chapter 3, the process of catalytic cracking involves the regeneration of catalyst particles by burning off the coke that forms on them in the reaction vessel. In the refinery process described in chapter 3 the quantity of this coke formed per barrel of gas oil feed varies between about 18 lb and 28.8 lb. The combustion of this coke creates very large

[32] Based on data in R. C. Ewing, "Activated Carbon Aids Water Treatment," *Oil and Gas Journal*, vol. 68, no. 18 (4 May 1970), pp. 134–35; and E. G. Paulson, "Activated Carbon Cleans Effluent," ibid., vol. 68, no. 26 (29 June 1970), pp. 85–87. In the model's process, carbon is ignored—about 2 lb/1,000 gal—which cannot be regenerated and hence becomes a residual.

[33] The corresponding cost of withdrawal and pretreatment for fresh boiler feed is assumed to be $0.15/1,000 gal.

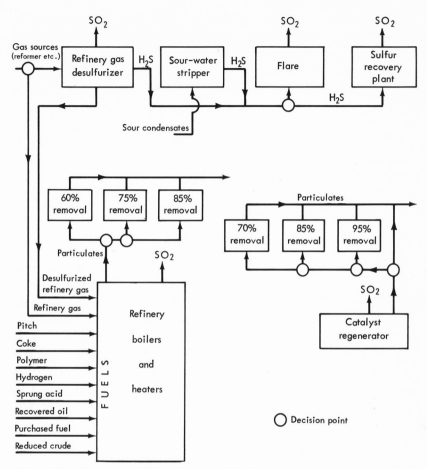

Figure 13.   Gaseous and other airborne residuals.

quantities of very hot gases, mainly CO_2 and CO.[34] (Regenerator outlet temperatures are on the order of 1,100°F.) Entrained in these gases are small particles of catalyst, some soot, and SO_2 formed from the sulfur in the coke. There may also be ammonia, aldehydes, oxides of nitrogen,

[34] Volumes of stack gas vary between 4.18×10^3 and 6.80×10^3 acf per bbl of fresh feed. They are estimated on the basis of pure carbon coke being oxidized to CO_2 and CO in the ratio of 1.1 to 1.0 with resulting gas at 1,100°F. Based on information supplied in the Bechtel technical review.

See also J. G. Wilson and D. W. Miller, "The Removal of Particulate Matter from Fluid-Bed Catalytic Cracking Unit Stack Gases," *Journal of the Air Pollution Control Federation*, vol. 17, no. 10 (October 1967), pp. 682–85; and U.S. Public Health Service, *Atmospheric Emissions from Petroleum Refineries: A Guide for Measurement and Control*, PHS Pub. 763 (1960), pp. 20 and 35.

cyanides and carbon monoxide, but the reduced list is reflected in the model.

Because of the value of the catalyst, over 99 percent of the particulates are removed from the gases within the regenerator unit as a matter of standard practice.[35] These catalyst "fines" rejoin the catalyst stream. The concern here is with the very small percentage of particulates that leave the regenerator vessel in the stack gases. These are assumed to amount to 0.0526 lb/1,000 actual cubic feet of stack gas or between 0.22 lb and 0.36 lb/bbl of fresh feed. Table 13 summarizes the residuals generation for each cracking process per barrel of fresh feed.

The particulates leaving the regenerator may be discharged to the atmosphere, or the stack gases may be passed through one of three alternative particulate removal processes (see figure 13): a low-efficiency dry cyclone removing 70 percent of the particulates; a somewhat more efficient cyclone giving 85 percent removal efficiency; or an electrostatic precipitator giving 95 percent removal.[36] (The SO_2 residual is assumed not to be affected by the discharge modification processes chosen.) The costs per 1,000 actual cubic feet of stack gas are respectively $0.023, $0.034, and $0.090.[37] The particulates removed from the gases become a solid residual that must be trucked away or landfilled on the refinery site. The cost of disposal of these solids is assumed to be $2/ton.

Provision of Refinery Heat Requirements

When fuel is burned to provide the necessary heat inputs, both directly to petroleum streams and to water for steam, some residuals are generated. If the fuel has a significant sulfur content, one residual will be SO_2. The burning of coke (from the coker); pitch (from the H-oil unit); purchased residual fuel oil; reduced crude (from the vacuum fractioning tower); and recovered oil (from the API separator unit) will produce particulates as well as SO_2. In most cases, of course, the refiner is free to burn or not to burn the particular fuel (this is not true of sprung acid or

[35] See Wilson and Miller, "Removal of Particulate Matter."

[36] These efficiencies are not broken down by particle size, though this would be desirable, since an aggregate efficiency of removal of 90 percent generally means that all large particles are being removed but that far more than 10 percent of the dangerous small particles are entering the atmosphere.

[37] Costs and removal efficiencies are based on data in U.S. Department of Health, Education, and Welfare, NAPCA, *Control Techniques for Particulate Air Pollutants*, Pub. AP-51 (January 1969); and API, *Removal of Particulate Matter from Gaseous Wastes: Wet Collectors* (1961), especially sec. 4 and table 8.

Table 13. Airborne Residuals Generation in Catalyst Regenerators

Residuals	E. Texas VGO, 66.9% conv.	E. Texas VGO, 83.5% conv.	Arabian VGO, 66.9% conv.	Arabian VGO, 83.5% conv.	E. Texas KGO, 66.9% conv.	E. Texas KGO, 83.5% conv.	Arabian KGO, 66.9% conv.	Arabian KGO, 83.5% conv.
SO_2 (lb/bbl)	0.30	0.56	1.44	2.68	1.28	1.88	3.10	4.44
Flue gas (1,000 acf/bbl)[a]	4.18	6.04	4.18	6.04	4.47	6.60	4.47	6.80
Particulates								
(lb/1,000 acf)	0.0526	0.0526	0.0526	0.0526	0.0526	0.0526	0.0526	0.0526
(lb/bbl)	0.220	0.318	0.220	0.318	0.236	0.348	0.236	0.358

Note: VGO = virgin gas oil; KGO = coker gas oil.
[a] Actual cubic feet at 1,100°F, regenerator outlet temperature.

recovered oil sludge from the API separator, which, in the model, must be burned); and in the case of purchased heat input, the model offers a choice between 0.5 percent, 1 percent, and 2 percent sulfur fuel so that a major part of the generation of sulfur oxide is subject to some control. In addition, for each major fuel (residual fuel oil, petroleum coke, and pitch), add-on alternatives for the removal of particulates have been included. These treatment units are all assumed to be dry cyclones and three alternative efficiencies are available: 60 percent, 75 percent, and 85 percent removal.[38] Their costs were calculated from the formulas reported in the Implementation Planning Program developed for the Environmental Protection Agency.[39] The argument for each of the cost functions (capital plus the elements of operating costs for each efficiency) is size in actual cubic feet per minute (acfm), and it has been assumed that requirements will be met by replication of units producing 100,000 acfm of the gases. This assumption allows the calculation of average total costs per 10^3 acfm and then the translation of these into costs per pound of particulates flowing into the unit by using assumed flue gas loadings for each fuel.

Table 14 summarizes the assumptions concerning residuals generation from fuel combustion at the refinery, and table 15 shows the costs per pound of particulates entering each of the stack gas-cleaning units.

SOLID RESIDUALS

There are only two solid residuals explicitly dealt with in the refinery model: the collected particulates from the modification processes attached to the catalyst regenerator and boilers and the ash from sludge incineration. A cost of $2/ton for the disposal of both of these residuals is assumed.

This completes the description of the structure of the petroleum refinery model, the sources of data, and the assumptions and calculations behind the linear programming vectors. The next chapter contains the first illustrations of the application of the model.

---

[38] Only the first two are available for the oil combustion residuals.

[39] See TRW Systems Group, *Air Quality Implementation Planning Program*, prepared for EPA (November 1970), vols. 1 and 2 (also available as PB 198 299 and PB 198 300 respectively from NTIS).

Table 14. Residuals from the Provision of Refinery Heat Input

	Sulfur weight (*percent*)	Heat content (10^6 *Btu/lb*)	Heat content (10^6 *Btu/bbl*)	SO$_2$ emissions ($lb/10^6$ *Btu*)	Particulate emissions[a] ($lb/10^6$ *Btu*)
Purchased heat					
Very low sulfur	0.5	—	6.20	0.55	0.0545
Low sulfur	1	—	6.20	1.10	0.0545
High sulfur	2	—	6.20	2.20	0.0545
Reformer hydrogen	0	0.0331	—	0	0
Coker gas (E. Tex. crude)	1.85	0.022	—	1.68	0
Coker gas (Arabian mix)	3.40	0.022	—	3.19	0
Cat cracker gas, VGO					
66.9% conv. E. Tex. crude	1.41	0.022	—	1.27	0
83.5% conv. E. Tex. crude	1.04	0.022	—	0.94	0
66.9% conv. Arabian	5.11	0.022	—	4.65	0
83.5% conv. Arabian	4.20	0.022	—	3.82	0
Cat cracker gas, KGO					
66.9% conv. ETC	4.61	0.022	—	4.20	0
83.5% conv. ETC	3.14	0.022	—	2.86	0
66.9% conv. Arabian	12.2	0.022	—	11.1	0
83.5% conv. Arabian	9.15	0.022	—	8.31	0
Desulfurized refinery gas	0.1	0.025	—	0.08	0
Sweet coke (E. Tex. crude)	1.57	0.015	—	2.10	2.00
Sour coke (Arabian)	3.32	0.014	—	4.74	2.00
Polymer (from reformer)	0	—	6.45	0	0
Sprung acid	22.4	0.0088	—	51.0	0
Reduced crude, E. Tex. crude	1.28[b]	—	6.20	1.40	0.0545
Reduced crude, Arabian[b]	2.96[b]	—	6.20	3.29	0.0545
Recovered oil from API separator	1	0.01	—	2.00	0.053
H-oil gas, E. Tex. crude					
Low conv.	18.0	0.022	—	16.3	0
High conv.	13.3	0.022	—	12.1	0
H-oil gas, Arabian					
Low conv.	43.7	0.011	—	79.4	0
High conv.	32.4	0.015	—	43.8	0
Hydrocracking gas, max. jet fuel, Arabian	11.2	0.022	—	10.2	0
Pitch 1, from H-oil unit, E. Tex. crude	0.5	—	6.44	0.575	0.80
Pitch 2, Arabian crude	1.17	—	6.44	1.35	0.80

Source: Heating values generally taken from W. L. Nelson, *Guide to Refinery Operating Costs* (Tulsa, Okla.: Petroleum Publications, 1970), p. 4.

Note: VGO = virgin gas oil; KGO = coker gas oil.

[a] Particulate emission factors assumed to be sweet coke and sour coke, 0.03 lb/lb; pitch, 0.015 lb/lb; purchased residual fuel oil, 8 lb/1,000 gal. All other heat sources assumed to produce no particulates.

[b] See table 1.

TABLE 15.   Refinery Gas-Cleaning Design Information

	Coke	Residual fuel oil	Pitch
Stack gas generation rate (*acf*)	935,000[a]	161,000[b]	128,000[b]
Particulate loading (*lb/10³ acf*)	0.102	0.002	0.040
Cost/lb of particulates into cyclones			
High (85%) removal efficiency (*$*)	0.0040	[c]	0.0103
Med. (75%) removal efficiency (*$*)	0.0033	0.163	0.0086
Low (60%) removal efficiency (*$*)	0.0025	0.125	0.0066

Note: Assumed rate of gas flow for each unit, 100,000 acfm.
[a] Per ton.
[b] Per barrel.
[c] Not included.

V

THE BENCHMARK CASE:
SOME RESULTS AND COMPARISONS

The data about costs, processes, output quality, and residuals generation summarized in the last two chapters have been organized in the
linear programming framework described in chapter 2. The resulting
model has been solved repeatedly using different sets of prices and costs,
subject to different constraints on activity levels and with different sets
of process alternatives available in the A-matrix. The following three
chapters will detail these experiments with the model. This chapter concentrates on the results for a benchmark formulation to which later results may be compared.

The purpose of devoting a chapter to a single solution is twofold. First,
by going into some detail it may be possible to improve the reader's "feel"
for the model and his understanding of the changes implied by the new
prices and constraints introduced below. Second, in this chapter it will be
possible to pause at various points to make some comparisons between
the basic results and similar figures reported elsewhere. These comparisons, I hope, will provide some ground on which the reader may judge
how seriously to take the later results for which comparable figures are
unavailable.

It is appropriate to repeat at this point that the major purpose of this
monograph is to give a demonstration of a method, of a model type,
which I believe can be of great utility in regional environmental policy
formulation. The technical coefficients reflected in the results could un-
doubtedly be improved by knowledgeable petroleum refinery process
engineers, and I hope, too, that my success will be judged primarily on

the flexibility and potential of the model rather than on the accuracy of specific numbers.

A reasonable place to start the comparisons is with the set of process units to be included, at least potentially, in the benchmark refinery. Since there are two important sources of information on residuals generation, it will be useful to compare the benchmark refinery with the ones in these sources—studies by the American Petroleum Institute and the Federal Water Pollution Control Administration.[1] In particular, how complete and representative is the refinery if the aim is to model a standard gasoline refinery without lube oil processing or petrochemicals manufacturing?[2]

Unfortunately the API Study contains a rather cryptic description of the five categories into which the surveyed refineries are divided. These categories are:

A. Topping.
B. Topping and cracking.
C. Topping, cracking, and petrochemicals.
D. Integrated (topping and cracking plus lube processing).
E. Integrated and petrochemicals.[3]

Since I have not included lube oil processing or petrochemicals manufacture, it appears that the model refinery has to be considered as falling in category B, and it must be assumed that this category includes plants performing the whole range of processes aimed at increasing gasoline yield per barrel of crude. This assumption seems reasonable in light of the makeup of the questionnaire on which the study was based.[4] On the inventory sheet of refinery facilities on the questionnaire, the refiner is supposed to indicate which of the following types of facilities he has (and also to give specific type and capacity): crude distillation, vacuum distillation (together these constitute topping), thermal processes, catalytic cracking, catalytic reforming, alkylation, polymerization, coking, as-

---

[1] API, Committee for Air and Water Conservation, *1967 Domestic Refinery Effluent Profile* (September 1968); and U.S. Department of the Interior, FWPCA, *The Cost of Clean Water*, vol. 3, *Industrial Waste Profile No. 5*, "Petroleum Refining" (November 1967). The first will be referred to as the API Study and the second as the FWPCA Study.

[2] In an actual situation it may be difficult to determine at what point petrochemicals manufacture begins, but the model refinery stops with by-product streams identified as feed sold for petrochemicals, with the exception of the inclusion of off-gas desulfurization required before such sale.

[3] API Study, p. 1.

[4] Ibid., pp. 56–63.

phalt, asphalt blowing, lubes, petrochemicals (including five subgroups), and product treating and finishing. Since the topping processes are easily identified, and since lubes and petrochemicals appear explicitly, it seems clear that "catalytic cracking" is used in the classification nomenclature as a surrogate for different sets of processes aimed primarily at gasoline manufacture; it does *not* indicate that class B refineries do only topping and cat cracking. The benchmark refinery includes processes from all the major subheads subsumed under "catalytic cracking" except polymerization and asphalt and asphalt blowing, and thus my results should be at least roughly comparable to the profile data.

The FWPCA Study is much more explicit and ambitious in defining the process mix behind its estimates of residuals generation. For the FWPCA "typical technology" refinery the processes included are crude oil storage; electrostatic desalting; atmospheric and vacuum distillation; viscosity breaking or coking; catalytic reforming (platforming); fluid catalytic cracking; hydrotreating (of naphtha, some gas oil, heavy distillate, and deasphalted reduced crude); polymerization; sulfuric acid alkylation; solvent refining of aromatics (from the reformer); solvent dewaxing; propane deasphalting; drying and sweetening of various streams; three lube-oil-finishing steps; and product-blending storage and packaging. In the FWPCA Study an "older" and a "newer" technology are also defined. The important difference between typical and older technology is that the latter has no coker or hydrotreating units. The new technology has hydrocracking and processes of the H-oil type. Hence the newer technology is comparable to my "advanced" refinery.[5]

The basic refinery model lacks the following from the list above: polymerization, solvent refining, solvent dewaxing, propane deasphalting, the lube oil processes, and the product-handling stages.[6] But because the FWPCA Study did not, in fact, develop residuals generation data for many of the processes on which its refinery definitions are based, my results should be more closely comparable than an inspection of flow charts would seem to indicate. Thus of the processes available in the FWPCA typical refinery but not in the basic model, deasphalting, the lube oil processes, and blending and packaging are not actually reflected in the FWPCA estimates of residuals generation because of lack of data. In addition, for crude oil and product storage and for solvent refining, residuals generation data are incomplete. The model, then, lacks poly-

[5] See FWPCA Study, figs. 1–3.
[6] I do not include a liquid residual contribution from crude oil storage.

merization and solvent dewaxing, which are fully reflected in the FWPCA results. Solvent refining and the various storage activities are not in the model but are only partially reflected in the FWPCA results. Both models effectively lack propane deasphalting and lube oil processing.[7]

Thus it has been established that the model is at least roughly comparable in structure to the refinery assumed by FWPCA in deriving estimates of some water pollution discharges and to the class B refineries whose discharges were surveyed by the API in 1967. I now turn to a detailed description of the model's solution in the benchmark case and to a comparison of the discharges in this benchmark case with those reported in the FWPCA and API studies. It will be convenient to summarize and discuss the benchmark results in five categories: input of crude, product mix, major processing units used, utilities, and residuals.

CRUDE OIL INPUT

In the base solution the model refinery charges 111,000 bbl/day of the East Texas crude (low sulfur) and 39,000 bbl/day of the Arabian mix crude (high sulfur). In setting up the problem, the price of the Arabian mix crude was kept low, approximating the world market price plus transportation from the Persian Gulf to the East Coast of the United States. This price is low enough that without additional constraints the refinery would charge only this higher-sulfur crude. In the base case the amount of the Arabian mix available is limited to 39,000 bbl/day, and in a subsequent set of runs, this constraint is relaxed to observe the effect of increasing the average sulfur content of the crude charged.[8] It should be noted, however, that the crudes also differ slightly in their fractioning characteristics (see table 1) so that in these later experiments I am not really holding everything but sulfur content constant. But the fact that the gasoline-blending products of higher-sulfur crude are generally lower octane may be attributed in a rough way to their higher sulfur content, and thus they may be assumed to fall within the desired ceteris paribus condition.

[7] See FWPCA Study, table 5.

[8] As I pointed out in chap. 3, I had originally intended to use this device to investigate the impact of import quotas. Even though this is no longer my intention, the reader may think of the upper limit on the Arabian mix as an import quota if he wishes.

PRODUCT MIX

The product prices and input costs used in the benchmark solution were summarized in chapter 3. The major products of the refinery are, of course, premium and regular gasoline. In the base solution, 35,100 bbl of 100-octane premium and 51,150 bbl of 94-octane regular are produced. The average tetraethyl lead contents are 2.50 cubic centimeters per gallon (cc/gal) and 2.46 cc/gal respectively. This total gasoline production was chosen to agree with the production figures used by Dunmyer et al., which in turn were based on average production rates reported by the Bureau of Mines for 1969.[9]

The other products in the benchmark solution are shown in table 16, along with a comparison of this product mix with that of Dunmyer et al. and a materials balance. The basic solution differs from the Dunmyer study and hence from 1969 Bureau of Mines data, principally in the breakdown of liquid products between kerosene-diesel oil and straight-run gasoline. My production of gasoline has been chosen to match theirs; my production of distillate and residual fuel oil has been made as large as the particular crudes would allow while still leaving some flexibility in the model.[10] It will be seen that in the advanced technology refinery some additional flexibility is gained, and kerosene yield can be significantly increased.

MAJOR PROCESSING UNITS USED

In the basic solution only the isomerization unit among the major available units is not chosen for use.[11] In order to facilitate further comparison with the Dunmyer study and with the FWPCA Study, table 17

[9] J. C. Dunmyer, Jr., R. E. Froelich, and J. L. Putnam, "The Cost of Replacing Leaded Octanes" (paper presented at the 36th Midyear Meeting, API, Division of Refining, San Francisco, 13 May 1971), table 7; and U.S. Bureau of Mines, "Crude Petroleum, Petroleum Products and Natural-Gas-Liquids: 1969 Final Summary," *Mineral Industry Surveys*, 15 December 1970.

[10] Thus, for example, distillate fuel oil production could be pushed up to 17,600 bbl/day, but the shadow prices on gasoline production would be about $25/bbl and on distillate, about $17/bbl. This indicates the extent to which the model is being "strained." I say more about shadow prices below.

[11] This appears reasonable. See, for example, Dunmyer, Froelich, and Putnam, "Cost of Replacing Leaded Octanes," where butane isomerization is never chosen under any lead constraint. In the benchmark refinery, sufficient isobutane is obtainable from the catalytic crackers to meet the input needs of the alkylation unit. I do allow the purchase of 600,000 lb (3,000 bbl/day) of normal butane. This agrees with Dunmyer, Froelich, and Putnam.

TABLE 16.   Benchmark Solution Product Mix with Materials Balance and
Comparative Figures

	Quantity produced per day in standard units	% of crude volume	% of crude volume, Dunmyer et al. product mix[a]	Weight (10^6 lb)
Products sold				
Refinery gas	2.944×10^6 lb			2.94
Kerosene-diesel oil	15,760 bbl	10.5	20.0[b]	4.53
Distillate fuel oil	17,400 bbl	11.6	15.0[c]	5.31
Low sulfur	8,880 bbl			
Medium sulfur	8,230 bbl			
High sulfur	290 bbl			
Polymer	660 bbl			0.21
Recovered sulfur	40.0 LT			0.09
Premium gasoline	35,100 bbl	22.4	22.4	9.38
Regular gasoline	51,150 bbl	34.1	33.5	13.67
Residual fuel oil	3,000 bbl	2.0	2.8	1.01
Straight-run gasoline sold as petrochemical feed	16,360 bbl	10.9	0	4.09
Subtotal: products sold				41.23
Products used internally				
Hydrogen (burned)	100,250 lb			0.10
Sweet coke (burned)	1,180,000 lb			1.18
Sour coke (burned)	260,000 lb			0.26
Coke burned in catalyst regeneration				1.54
Subtotal: products used internally				3.08
Residuals				
SO$_2$ as S (corrected for sulfur in cokes burned and in fuel purchased)[d]				0.01
Other residuals (H$_2$S, oil, phenols, NH$_3$, ½ of BOD)[e]				0.02
Subtotal: residuals				0.03
Miscellaneous				
Isobutane (price taken as 0 for net production)				0.17
Total material output				44.51
Inputs				
Crude charged (150,000 bbl total)				43.92
Normal butane purchased				0.60
Total material input[d]				44.52

[a] J. C. Dunmyer, Jr., R. E. Froelich, and J. L. Putnam, "The Cost of Replacing Leaded Octane" (paper presented at the 36th Midyear Meeting, API, Division of Refining, San Francisco, 13 May 1971).

[b] This figure is the sum of the categories "kerosene-jet fuel A-1" and "diesel oil" in the Dunmyer study.

[c] This is the production of No. 2 fuel oil that is comparable to what I have been calling distillate fuel oil.

[d] I exclude purchased fuel in this balance.

[e] I assume as a rough approximation that actual material residual from refinery inputs equals one-half of the weight of oxygen measured in BOD.

shows the size of each major unit in relation to the total crude charged for each study and for the benchmark solution. (Some of these units, such as hydrotreating, represent totals for a type of process that, in the model, is applied separately to different petroleum streams.) Since the model represents a grass roots design problem, the quantity of a process used in the solution may be taken as an indication of the necessary size of that unit in the refinery (before any correction for variations in daily use, safety, outage factors, etc.).

Table 17 indicates that in terms of process units used the basic refinery is quite similar to that postulated by FWPCA in its residuals study. The only significant differences between these two are in hydrotreating and catalytic reforming, for both of which units the FWPCA Study shows much larger use. In the case of hydrotreating this is because it is assumed in the FWPCA Study that some cat cracker feed will be hydrotreated, while in the basic solution only reformer feed goes through this process. The difference in reformer sizes is due simply to a difference in the proportion of the crude assumed to be in the naphtha cut and hence appropriate for reforming. As the fourth column shows, the FWPCA typical refinery is closer overall to the average U.S. refinery (reflecting all refinery types and sizes) than is the benchmark solution.

TABLE 17. Major Process Units in the Benchmark Solution as a Proportion of Crude Charged

Process unit	Size (*bbl/bbl of crude charged*)			
	Basic model	Dunmyer et al.[a]	FWPCA typical refinery	Average U.S. refinery, *Oil and Gas Journal* survey[b]
Atmospheric distillation[c]	1.00	0.952	1.00	1.0
Coking	0.133	0.061	0.150	n.a.
Hydrotreating	0.139	0.199	0.350	0.287
Reforming	0.139	0.199	0.200	0.217
Catalytic cracking	0.466	0.355	0.500	0.340
Alkylation	0.076	0.075	0.060	0.058
Sweetening	0.393	n.i.	0.500	n.a.

Note: n.i. = not included; n.a. = not available.

[a] J. C. Dunmyer, Jr., R. E. Froelich, and J. L. Putnam, "The Cost of Replacing Leaded Octane" (paper presented at the 36th Midyear Meeting, API, Division of Refining, San Francisco, 13 May 1971), table 11.

[b] "1971 Survey of U.S. Refineries," *Oil and Gas Journal,* vol. 69, no. 12 (22 March 1971), p. 94.

[c] Vacuum distillation as a fraction of crude charge will vary with crude charged. It is not available in the model's solution. For the United States as a whole, vacuum distillation capacity is 35.7 percent of atmospheric distillation capacity.

The differences between the base solution and the Dunmyer study are the use of coking, reforming, and catalytic cracking. These differences are quite consistent with what has already been observed about product mixes. The coker use differs because of the different assumptions about the size of the reduced crude fraction from the vacuum tower; the cat cracker use differs because of assumptions about gas oil cuts. In general the Dunmyer study reflects a crude that has larger naphtha and kerosene-diesel oil cuts, and smaller gas-oil and reduced crude fractions. The naphtha fraction is particularly important because it is the major reformer feed, and hence the major input to a process producing a very high clear-octane blending stock.

Perhaps, however, what is most striking about table 17 is how similar the refineries are. One may see this as evidence that the basic model is not outrageously wrong; or, more cynically, as evidence that a reasonable approximation to the basic solution could have been obtained without using a linear program at all. But this interpretation misses the point. However easy it might be to put together a set of process units that approximate recent average experience, it is certainly not easy to predict how refinery configuration will change in response to combinations of such external influences as stricter residuals discharge controls, proposed changes in gasoline lead levels, and rapidly increasing demand for jet (turbine) fuel. To do this a tool is needed that reflects the extremely complex set of technological possibilities and constraints facing the refiner and that relates these, in turn, to some approximation of his decision rule. This the linear programming model form is admirably equipped to do.

Utilities

In the model, electric power purchases have been subsumed into the activity objective function entries. This leaves fresh heat inputs, water withdrawals, and steam generation in the "utilities" category.[12] As for fresh heat, in the benchmark solution $33,900 \times 10^6$ Btu are purchased from outside. It is assumed that this is in the form of residual fuel oil, and, as discussed in chapter 4, that 0.5 percent, 1 percent, and 2 percent sulfur grades are all available. (In this solution the cheapest, highest-sulfur

[12] I have assumed that steam is used only for processing, i.e., that pumps, compressors, etc., are driven by electricity.

TABLE 18.　In-Refinery Sources of Fresh Heat

	Quantity burned (*lb*)	Heat supplied (*10^6 Btu*)
Excess hydrogen (from reformer)	100,000	3,310
High-sulfur coke	260,000	3,840
Low-sulfur coke	1,180,000	17,570
Sprung acid (from spent caustic neutralization)	15,500	140
Recovered oil (from API separator)	1,000	10
Total from in-refinery sources		24,870 (42.3%)
Total purchased heat		33,900 (57.6%)
Total fresh heat applied		58,770 (99.9%)

grade is purchased.) It is assumed that the heat content of this purchased residual is 6.20×10^6 Btu/bbl, and total purchases then amount to 5,475 bbl/day.

Additional heat is provided within the refinery by burning various by-products and recovered substances. These additions are shown in table 18. Again, using the assumed heat content of residual fuel oil, 6.2×10^6 Btu/bbl, I can convert total refinery fresh heat use into fuel oil equivalents. The $58,770 \times 10^6$ Btu/day are equivalent to 9,480 bbl/day of residual fuel oil, or about 6.3 percent of crude throughput (by volume). This may be compared with a rule of thumb used in the industry—that refinery heat requirements should not exceed 10 percent of crude throughput on this basis.[13]

Process uses of steam in the basic solution require the production of 22.2×10^6 lb of steam per day, or 148 lb/bbl of crude charged. Of this total about 55 percent is produced in the refinery boilers burning purchased or internally produced fuel, and 45 percent is produced by recovery of heat generated, as in catalyst regeneration, in other processes. I have assumed, in effect, that about 73.5 percent of this steam is condensed within the processing units, becoming contaminated with H_2S, phenols, oil, and—in consequence of the total organic load—BOD.[14]

The model determines separately the quantities of water withdrawn for three different purposes: cooling, production of steam, and desalting.

[13] Letter to the author from Lionel S. Galstaun of Bechtel Corp., 11 May 1972.
[14] See chap. 3.

I distinguish between these three streams because the water for each is assumed to require different degrees of treatment before use (reflected, for simplicity, only in the objective function) and because, as a result, recirculation to each use arises as an alternative after a different level of effluent treatment. The quantities withdrawn in the basic solution per day are:

For cooling (assuming a 35°F rise in temperature)	154,350 × 10^3 gal
For the desalter	360 × 10^3 gal
For steam production	2,662 × 10^3 gal

Notice that cooling water withdrawals are about 98 percent of total withdrawals. This may be compared with data from studies on water use in manufacturing in the *Census of Manufactures, 1963*.[15] On the average across the United States, in Standard Industrial Classification (SIC) code 2911 (petroleum refining) cooling withdrawals amounted to 87 percent of total withdrawals. Considerable recirculation, however, is reflected in the corresponding gross water use figures, without being split up between cooling and process categories. Thus it is impossible to obtain a figure exactly comparable to the ratio from the benchmark solution in which no recirculation of any stream is being practiced.

Succeeding chapters will deal with the costs (or gains) implied for the refiner by the imposition of new product quality or quantity requirements and of effluent charges or constraints. In addition it will be possible to observe the impact on costs of the availability of new technology, the hydrogen-intensive cracking processes. It is therefore appropriate as part of the stage-setting function of this chapter to include some discussion of benchmark costs, both average and "marginal."[16]

The marginal costs of interest here are the shadow prices (or dual values) associated with the product constraints. These are summarized below and compared with ex-refinery prices for these same products in the New York–Philadelphia area of the East Coast.[17]

---

[15] U.S. Bureau of the Census, *Census of Manufactures, 1963*, vol. 1, *Summary and Subject Statistics*, chap. 10, "Water Use in Manufacturing" (1966), pp. 10–24, 10–25. This source indicates that gross use in petroleum refining was 4.4 times as large as withdrawals for the year of the study.

[16] In a linear model, of course, "marginal" has a special meaning because of the discontinuities in the first derivatives. For simplicity I identify shadow prices on constraints as marginal costs without qualification. See, however, the discussion in chap. 7.

[17] Major price sources were *Platt's Oilgram Price Service, Oil and Gas Journal*, and the Bechtel Corporation's technical review of an earlier draft of this study.

	Benchmark shadow price	Ex-refinery price (1972)
	(dollars)	
Regular gasoline	5.54	5.38–5.59
Premium gasoline	6.00	6.22–6.43
Kerosene	0[18]	4.96
Distillate fuel oil	3.87	4.60
Residual fuel oil	4.31	3.10

The shadow prices reflect the cost of crude oil, total refining costs, and the opportunity cost of the priced by-products, but they do not reflect any allowance for sales or administrative overhead. It is difficult, of course, to know how the actual ex-refinery prices are arrived at because of the oligopolistic nature of the petroleum industry, but in a more competitive world one would hope to find marginal costs in the model below the ex-refinery prices by some small amount. The actual pattern observed is not quite that neat. The differential between premium and regular shadow prices appears to be too small. And the relative prices of distillate and residual fuel oils is reversed. In the actual East Coast markets, however, the residual fuel oil price is strongly affected by import policies that allow the entry of large quantities of Caribbean residual. These policies are reflected in turn in the processing regimes and costs in Caribbean refineries, which tend to turn out much larger residual fractions than mainland refineries.[19] Hence the shadow prices at that end of the scale cannot really be expected to match the market prices.[20] Overall it can at least be said that the marginal costs facing the refinery in the base case are of the proper order of magnitude.

Average refinery costs are shown in table 19 under five headings: purchased inputs (crude oil and butane); total process unit costs exclusive of heat, steam, and cooling services (but inclusive of electricity); utilities

[18] The basic technology model can only change its kerosene production by changing its crude mix. The lower limit on production in the benchmark case is set at 15,000 bbl/day, less than the kerosene produced by the benchmark crude mix and just equal to the amount available from a 100 percent charge of the East Texas crude.

[19] See, for example, Hittman Associates, *Study of the Future of Low Sulfur Oil for Electrical Utilities*, prepared for EPA (February 1972), available from NTIS as PB 209 257.

[20] An additional problem concerns distillate fuel oil. The average sulfur content of the benchmark distillate is 1.3 percent, too high to meet accepted standards. Were sulfur content limited to 0.3 percent, the shadow price in the model would increase to $6.45/bbl at 17,100 bbl/day, the largest production of this grade of distillate consistent with the other constraints. (There is some volume loss in hydrotreating, hence the lower volume of low-sulfur distillate fuel oil attainable.)

TABLE 19.   Average Refinery Costs and Receipts

	Costs/bbl of crude ($)	Percent of total average cost	Receipts/bbl of crude ($)	Percent of total receipts
Purchased inputs				
(crude oil + n-butane)	3.636	82.3		
Process units exclusive of utilities	0.505	11.4		
Utilities	0.134	3.0		
Gasoline blending	0.131	3.0		
Miscellaneous	0.013	0.3		
Total average cost	4.419	100.0		
Products and by-products sold in model			0.940	17.6
Products constrained valued at approx. ex-refinery prices				
Premium gasoline ($6.22/bbl)			1.458	27.2
Regular gasoline ($5.38/bbl)			1.834	34.3
Kerosene ($4.96/bbl)			0.524	9.8
Distillate fuel oil ($4.60/bbl)			0.534	10.0
Residual fuel oil ($3.10/bbl)			0.062	1.1
Total receipts/bbl crude			5.352	100.0
Net income/bbl, or 17.4% of sales			0.933	

(heat, steam, and cooling water); gasoline blending (cost of TEL additions); and miscellaneous (including gas desulfurization and gasoline scrubbing). I also show refinery receipts per barrel of crude, both those from sales activities built into the model (such as recovered sulfur, refinery gases, etc.) and those estimated for constrained products on the basis of ex-refinery prices.

By far the largest cost item is crude oil, which with the small amount of purchased normal butane accounts for over 80 percent of total costs. The next largest item, process units, is about one-eighth as important. It is perhaps surprising to find that the costs of TEL additions are about the same as total refinery utility costs (exclusive of electricity). On the receipt side, total gasoline production valued at ex-refinery prices accounts for over 60 percent of the total. Kerosene and distillate fuel oil make up another 20 percent, and all the rest of the items sold amount to another 20 percent.[21] Under the price assumptions and production constraints of the model, net income for the refinery is about 17.4 percent of

[21] This "all-other" category is dominated by refinery gas sales, which in the benchmark amount to $0.428/bbl of crude, or about 8 percent of receipts.

sales. It should be emphasized that this figure is not directly comparable to the reported earnings of petroleum firms but is merely provided as part of the background of the model.

RESIDUALS

The basic purpose of the model is to predict how residuals discharges from a petroleum refinery will vary with changes in various direct policies and underlying conditions. The discharges from the benchmark model are determined under the assumption that no public actions have been taken to regulate or to price discharges of SO_2, particulates, BOD, NH_3, or phenols. On the other hand, it has been assumed that the refiner faces some kind of pressure, either in the form of a law or simply of aroused public opinion, which forces him to install API oil-water separators and sour-water strippers rather than to discharge the quantities of oil and dissolved hydrogen sulfide residuals generated in production.[22] The total cost of this equipment and its operation per barrel of crude is \$0.0047/bbl. In chapters 7 and 8, when the impacts of explicit public policies aimed at reducing residuals discharges are discussed, only costs in excess of this figure will be included. In table 20, I summarize the base discharge levels for the residuals handled in the model. In addition, I include figures from other studies for comparable refineries and levels of treatment (where given).

Among the waterborne residuals, the benchmark results appear to be low for sulfide and high for phenols and to fall within the range of reported discharges for oil, BOD, and NH_3.[23] On the air side, the results are compared with the average estimated discharges per barrel from six major refineries in the Delaware estuary region as reported in the Phila-

[22] Recall that my definition of residuals generation involved measurement downstream from any equipment installed because of its profitability to the refiner in the absence of any residuals discharge controls, formal or informal. Thus the generation of particulates in catalyst regeneration is measured after considerable effort has been expended to trap and recirculate over 99 percent of the catalyst fines originally entrained in the combustion gases.

[23] As noted in table 20, the API Study does not explicitly state that sour-water stripping is employed by the refineries in the sample, but it appears this must be the case. If my refinery did not employ this modification process, discharges of H_2S dissolved in waste process water would be 100 times as large as they are and thus about 50 times as large as for the API sample average. Such an error seems highly unlikely given the methods used to estimate the H_2S residual loads from each process unit.

TABLE 20.   Residuals Discharges: Results for the Benchmark Solution
Compared with Other Data

Residuals per bbl of crude charged	Bench-mark solution	API Study[a]	FWPCA Study[b]	Beychok[c]	Delaware estuary refineries[d]
To water:					
Heat (10^6 *Btu*)	0.300	—	—	—	—
Sulfide (*lb*)	0.003	0.006	0.006	0.006	—
Phenols (*lb*)	0.032	0.013	0.016	0.010	0.012
Oil (*lb*)	0.047	0.025	—	0.070	—
BOD (*lb*)	0.060	0.073	0.054	0.120	0.145
NH₃ (*lb*)	0.021	0.085	—	0.015	—
To air:					
SO₂ (*lb*)	1.429	—	—	—	0.985
Particulates (*lb*)	0.423	—	—	—	0.190

Note: Dashes = not available in study cited.

[a] From API Study, p. 25, table 2, weighted average discharges/bbl crude for 23 class B refineries reporting discharge after "primary" treatment. (Primary treatment includes gravity oil separation and similar processes and must include sour-water stripping, although this is not explicitly stated.)

[b] From FWPCA Study, table 6, average values for typical technology after API separator.

[c] M. R. Beychok, *Aqueous Wastes from Petroleum and Petrochemical Plants* (John Wiley, 1967), table 3A, pp. 44–46, based on data for post–API separator effluents.

[d] Particulate and SO₂ emissions based on "Metropolitan Philadelphia Interstate Air Quality Control Region Inventory of Emission Sources" (EPA, Division of Applied Technology, Air Pollution Control Office, Research Triangle, Durham, North Carolina, 1972, tape). Unweighted average of per barrel emissions corrected for particulate removal from combustion sources and uncorrected for process sources. BOD and phenol discharges based on 1968 discharges as measured by Delaware River Basin Commission.

delphia Air Quality Region Source Inventory.[24] According to this comparison both SO₂ and particulate emissions are high from the benchmark refinery. There appear to be two major reasons for these discrepancies. First, none of the refineries in the region report burning coke as in-house fuel,[25] while in the benchmark solution about 36 percent of total fresh heat (exclusive of heat captured in catalyst regeneration) is provided by coke. Twenty percent of this coke has a sulfur content of 3.3 percent, considerably above the average sulfur content of residual fuel

[24] "Metropolitan Philadelphia Interstate Air Quality Control Region Inventory of Emission Sources" (EPA, Division of Applied Technology, Air Pollution Control Office, Research Triangle, Durham, N.C., 1972, tape). There are seven large refineries in this region, but one is obviously under-reported in the inventory, and I do not include it in the average.

[25] Only two of the refineries have cokers according to the 1972 process-unit inventory reported in Ailleen Cantrell, "Annual Refining Survey," *Oil and Gas Journal*, 27 March 1972, pp. 135, 142, 150, 152. Two of the refineries report combustion processes that might involve coke, since the particulate emissions are too large to be accounted for by the fuel types and quantities found in the source inventory.

oil reported by the Delaware refineries. Thus the fuel mix in the benchmark solution is quite different from that reported, on the average, by the actual refineries used for comparison. Indeed the burning of the high-sulfur (2 percent) purchased residual and of both refinery cokes (2 percent and 3.3 percent sulfur respectively) accounts for over 60 percent of total SO_2 emissions and about 68 percent of total particulate emissions. If the refinery were required to burn 1 percent sulfur residual and if it were able to sell its sweet coke and make up the heat requirement again with 1 percent sulfur residual, the emissions per barrel would be 1.064 lb/bbl of SO_2 and 0.195 lb/bbl of particulates.

A second reason for the very high particulate loading in the model solution may be that my estimate of particulates per 10^6 Btu of coke is high. I have used 2 lb/10^6 Btu, or 0.03 lb/lb of coke, which roughly corresponds to a coal with a 4.5 percent ash content burned in a large industrial boiler. There should be very little ash in petroleum coke, and particulate emissions should consist mainly of unburned carbon, but no estimates of emissions per ton or per 10^6 Btu appear to be available. All this, of course, is based on the assumption that the inventory is accurate for the six refineries I used, which is a very big assumption.

Finally, it is interesting to note that recently the Environmental Protection Agency has taken DELMARVA Power Company to court because its generating plant on the Delaware estuary is burning very high-sulfur coke from the Getty refinery at Delaware City. The contract between the two firms apparently provides for a swap of coke for electricity. To the extent that similar arrangements occur elsewhere along the estuary, refinery emissions will tend to be lower and electrical generating station emissions higher.

Sulfur is a particularly interesting residual in this model, both because its sources are well defined and because some trouble has been taken to estimate the fate of sulfur at each process and treatment unit. It is therefore worth pausing to consider the refinery's sulfur balance as an example of how a major impurity in the input streams is translated into plant residuals, product impurities, and by-product by a complicated set of processes. Table 21 shows the sources and fates of sulfur in the benchmark solution. It is interesting to note that a very large fraction of the sulfur entering the refinery leaves as a product contaminant. If one considers only sulfur from the crude charged and allows for the fact that essentially all the sulfur in the purchased fuel goes up the boiler stacks as SO_2, this is even more striking, since of the 295,200 lb of sulfur in the crude over 45 percent is never removed from the petroleum despite the

TABLE 21.   Sources and Fates of Sulfur in the Benchmark Solution

Inputs and outputs	Pounds/day	Percent
Sulfur inputs		
In crude oils	295,200	88.8
In purchased residual fuel oil	37,200	11.2
Total sulfur input	332,400	100.0
Sulfur outputs		
Discharge of sulfur in SO_2	107,100	32.4
Discharge of H_2S in water	300	0.1
Residual sulfur (subtotal)	107,400	32.5
By-product sulfur recovered	89,700	27.0
Sulfur in refinery products	133,800	40.5[a]
	330,900[b]	100.0

[a] Product sulfur is based on the following assumptions: an average gasoline sulfur content of 0.08%; an average distillate fuel oil sulfur content of 1.3% (too high to match current standards for No. 2 fuel oil); plus product specifications already discussed in chap. 3.
[b] Approximately 1,500 lb unaccounted for.

many levels of processing and distillation. The fraction of crude sulfur recovered is about 0.30 and the fraction appearing as a residual is less than 0.24. As shown below, these fractions will change considerably when more desulfurized products (kerosene, distillate fuel oil, and residual fuel oil) are required.

It has been noted that the benchmark solution includes distillate and residual fuel oils of higher sulfur content than is generally acceptable. It is therefore worthwhile to pause and explore the effects of changing the product mix requirement to one in which both distillate and residual have sulfur contents of less than 0.5 percent. The first change would be in table 17 where hydrotreating as a proportion of crude still capacity increases to 0.210 to handle the quantities of distillate and residual that must be desulfurized. In table 18 total fresh heat input would increase by about 3 percent, and a slightly greater proportion would be from purchased residual (0.62 versus 0.58). The average cost net of by-products goes up by about $0.04/bbl to $3.52/bbl, but the marginal costs increase dramatically as evidenced by the shadow prices on the major product constraints, which go up by a factor of about 3.5.[26] The effect on residuals

[26] This increase serves to point out that meeting the lower-sulfur product is not simply a matter of desulfurizing every stream previously "sold" that is too high in sulfur. In particular I assume small volume losses in the relevant desulfurization activities so that more distillate and residual must be withdrawn from gasoline production. The blending stocks affected are generally high octane, so all the lead-octane constraints become binding.

discharges (table 20) is significant only for H_2S and NH_3, which increase
by 67 percent; and for BOD, which increases by 40 percent. These changes
are attributable to the introduction of the necessary hydrotreating process units. Finally, the fate of sulfur, as summarized for the benchmark
case in table 21, is considerably different under the changed product mix
constraints: by-product sulfur accounts for 48.6 percent versus 27 percent
of sulfur "output"; while sulfur in products declines to 14.7 percent of
this output.

The results discussed in this chapter correspond to a refinery that I will
refer to as the "basic" model; that is, a model without the hydrocracking
and H-oil process options discussed in chapter 3. When these two alternatives are included, the model will be referred to as "advanced." Table 22
summarizes the other conditions imposed on the basic refinery in arriving

TABLE 22.　Conditions Imposed on the Basic Refinery for the Benchmark Solution

Area	Condition imposed	Chapters containing variations
Technology	Hydrocracking and H-oil processes not available	6
Input availability	Cheaper, higher-sulfur crude limited to 39,000 bbl/day	6
Prices and costs	Cost of water withdrawals: 　Cooling, \$0.015/1,000 gal 　Desalter, \$0.070/1,000 gal 　Process steam, \$0.15/1,000 gal	6
	Cost of purchased fresh heat: 　0.5% sulfur, \$0.477/10^6 Btu 　1.0% sulfur, \$0.593/10^6 Btu 　2.0% sulfur, \$0.661/10^6 Btu	6
	Price of recovered sulfur: \$20/long ton	6
Product quality	Regular gasoline, octane $\geq$ 94; 　TEL $\leq$ 2.5 cc/gal	6
	Premium gasoline, octane $\geq$ 100; 　TEL $\leq$ 2.5 cc/gal	6
Product mix	Regular gasoline production 51,150 bbl/day Premium gasoline production 35,100 bbl/day	6, 9
	Kerosene production $\geq$ 5,000 bbl/day	6, 9
	Jet fuel production—none required	9
	Residual fuel oil production $\geq$ 1,000 bbl/day	6
Residuals discharge	Waste process water subject to oil-water separation and sour-water stripping	7
	Residual heat—no restrictions only	7
	SO_2 and particulates—no restrictions	8

at the benchmark solution. I do not repeat the entire list of product and input prices given in chapter 3, but only those that will be varied in the next chapter in exploring the sensitivity of the model and particularly of its residuals discharges to these underlying conditions.

VI

EFFECTS OF INDIRECT INFLUENCES ON THE GENERATION AND DISCHARGE OF RESIDUALS

Chapter 1 presented the argument that it is fundamentally misleading to take as constant coefficients relating industrial residuals generation to measures of sales, employment, or inputs. This is true for two important reasons. First, even if one is considering only the firm's response to outside action concerned directly with its residuals discharges (such as discharge standards), its options are not confined to the installation and operation of treatment facilities but extend back into its processing units and may involve changes in raw material quality, product quality, process type, and so forth. Thus if a firm's discharge of some residual is to be limited to X lb and its current generation is Y lb, there is no reason to assume that the response will include $Y - X$ lb of removal by "treatment." Rather, the firm may choose to make other changes, which result in $Y' < Y$ lb being generated, and to do less treatment. (This situation, of course, is significantly complicated by two further considerations: first these direct actions will frequently involve several residuals at once; and, second, they will generally result in the transformation of one residual into another.)[1]

Second, and perhaps even more important, industrial residuals generation, expressed as a ratio to output or input quantity, employment, and so forth, may change for reasons having nothing directly to do with residuals. Thus, product quality may be altered to respond to (or create)

---

[1] Only by-product production, or "recirculation" of the residual itself, actually removes residuals.

changing consumption patterns or to conform to government policies for consumer protection or other purposes. For example, the food-processing industry has shifted to more and more highly processed (residual-intensive) products over the last two decades. Or a new production process with new residuals generation characteristics may be developed and adopted to cut production costs or to allow the use of previously unused raw material.[2] Thus the kraft pulping process has come to dominate the paper industry because of its applicability to virtually any variety of wood.[3] Other government policies, such as those on international trade, may tend to shift raw material sources and hence composition, with implications for residuals generation. For example, the present oil import quota system tends to shift demand for crude oil to domestic producers.

The purpose of this chapter is to demonstrate the second of these two major points; that is, to show that residuals generated in a standard gasoline refinery vary with changes in product quality requirements, input prices, and raw material input quality. These influences on residuals generation are referred to as *indirect* to distinguish them from the actions taken by environmental quality management authorities and aimed directly at residuals discharges.[4] It is assumed throughout this investigation that the refinery is still in the design stage so that the fullest range of adjustment is always open. In keeping with my basically illustrative purpose, I do not examine every possible indirect influence, nor do I report on all possible combinations of those shifts I do include. But the results do cover some of the most significant indirect influences expected to impinge on petroleum refining over the next decade and thus are appropriate indicators of the magnitude of the shifts involved.

---

[2] The concern here is not with distinguishing between induced and exogenous technical change. However, to the extent that effluent charges or other policies make residuals discharge relatively more expensive one can expect to see a shift toward "residuals-generation-saving" technical change in the future.

[3] See Blair T. Bower, George O. G. Löf, and W. M. Hearon, "Residuals Generation in the Pulp and Paper Industry," *Natural Resources Journal*, vol. 11, no. 4 (October 1971), pp. 605–23; American Paper and Pulp Association, *Statistics of Paper—1964* (New York, August 1964); and American Paper Institute, *The Statistics of Paper: 1969 Supplement* (New York, December 1969).

[4] To avoid confusion, I include under direct actions only those designed to influence the discharges of the plant itself. Actions aimed at changing the characteristics or use of the plant's products for environmental quality reasons will be classed as indirect influences even though the same agency may be responsible for both policies. This distinction is most obviously illustrated by restrictions on the lead content of gasolines designed to change auto emissions. These restrictions I define as indirect influences from the point of view of the petroleum refinery's residuals problem.

CHANGING TECHNOLOGY

As pointed out in chapter 3, one of the most significant changes in petroleum refining technology, at least for the short-term future, is the advent of hydrogen-intensive processing units designed to crack a very wide range of charge stocks and to produce primary products in the gasoline to mid-distillate (kerosene) range. From the refiner's point of view, the major advantages of the new processes are high yields on a volume basis (often over 100 percent of the charge stock), flexibility in product yields (gasoline or kerosene-jet fuel may be stressed), and high product quality. This latter advantage arises because desulfurization (as well as denitrification and removal of certain metals) necessarily goes on simultaneously with the cracking reactions.

An obvious question, then, is how the basic refinery process configuration and residuals generation and discharge change when the hydrogen-intensive processes are available for use in the production of the benchmark product mix. Interestingly enough, the answer is that nothing changes; none of the new processes are adopted when the product mix includes leaded gasoline and the relatively small quantities of kerosene, distillate, and residual fuel oils chosen for the benchmark case.[5] Evidently flexibility and the higher yield of reformer feedstock are not worth the extra cost until the refiner is pushed to produce a more "difficult" product mix—for example, one in which kerosene production is greater than the amount naturally occurring in the crude, or in which no-lead, high-octane gasoline is required.[6] This will be seen below, but for now it is worthwhile observing that the hydrogen-intensive practices have not pushed out catalytic cracking in the real world either. According to the 1972 *Oil and Gas Journal* survey of all U.S. refineries, catalytic-cracking capacity measured on a fresh-feed basis was 34 percent of crude capacity,

[5] This result is quite insensitive to my estimates of the costs of hydrocracking. A reduction of costs to 59 percent of the estimated level is necessary to stimulate adoption of hydrocracking with the benchmark product mix.

[6] As currently set up, the model does not capture the value of flexibility—the ability to switch back and forth between major products—and to do so would require a dynamic model, or at least one in which the costs of building and operating the "switchable" unit would be compared with the cost of building two or more alternatives that would stand idle at least part of the time as product demands fluctuated. See, for example, George Stigler, "Production and Distribution in the Short Run," *Journal of Political Economy*, June 1939, pp. 305–27; and V. Kerry Smith and Joseph Seneca, "A Further Note on the Cost Implications of Fluctuating Demand," *Journal of Financial and Quantitative Analysis*, vol. 5, no. 3 (September 1970), pp. 369–76.

while hydrocracking capacity was only 6.4 percent.[7] Over the four years between the 1968 and 1972 surveys, additions to hydrocracking capacity in the United States have amounted to 432,000 bbl/stream day, while additions to cat-cracking capacity have been 394,400 bbl/stream day. The difference between the rates of addition has actually decreased over this period; between 1971 and 1972, 62,300 bbl/stream day of cat cracking and 63,700 bbl/stream day of hydrocracking capacity were added.[8] The ability of cat cracking to compete against hydrogen cracking is primarily attributable to the new higher-yield zeolite catalysts also developed in the late 1960s.[9] These catalysts are assumed in the cat-cracking vectors in the model. It appears, then, that it is quite reasonable for the model to reject hydrogen-cracking alternatives when faced with the "easy" benchmark product mix constraints.

CHANGING GASOLINE QUALITY REQUIREMENTS

Since one of the nation's major sources of atmospheric emissions (especially CO, NO_x, unburned hydrocarbons, and lead) is our fleet of private automobiles, it is not surprising that one of the great environmental quality debates has concerned how best to reduce these emissions. Although the rhetoric of the major parties involved has tended to obscure the issues, the federal standards now contemplated seem to require the installation of catalytic afterburner devices to reduce CO and hydrocarbon emissions. If this is true, the next question is: Can a catalyst be found that will not be subject to a very rapid loss of efficiency due to lead "poisoning," *or*, can gasolines be made without the addition of TEL (or with very small additions) to protect the catalysts? And this, of course, raises a second question: What octane level should be maintained in the gasoline?

It is unnecessary to trace the argument among auto manufacturers, petroleum refiners, and lead manufacturers as to the "best" course of

[7] Ailleen Cantrell, "Annual Refining Survey," *Oil and Gas Journal*, vol. 70, no. 13 (27 March 1972), p. 136.

[8] See the following issues of *Oil and Gas Journal:* vol. 69, no. 12 (22 March 1971), p. 94; vol. 68, no. 14 (6 April 1970), p. 116; vol. 67, no. 12 (24 March 1969), p. 116; and vol. 66, no. 14 (1 April 1968), p. 130.

[9] For an assessment of the trade-offs, see J. R. Murphy, M. R. Smith, and C. H. Viens, "Hydrocracking vs. Cat Cracking for Gas Oils in Today's Refineries," *Oil and Gas Journal*, vol. 63, no. 23 (8 June 1970), pp. 108–12. Their assessment: "Generally, cat cracking is now favored when gasoline is prime product; hydrocracking gets the nod for primary production of jet fuel or middle distillate."

action. Suffice it to say that all parties desire to hold down the extra costs they must incur and subsequently attempt to pass on to consumers. At the present writing the most likely outcome, at least for the next five years or so, appears to involve the introduction of catalytic converters on new automobiles, a reduction in the average lead content of gasolines sold (but probably not its complete elimination from all products), and a reduction in the average octane of gasoline accompanied by a reduction in compression ratios in new car engines.

The rules proposed by the Environmental Protection Agency (EPA) can be summarized as follows:

1. No gasoline may be sold by a refiner, wholesaler, or retailer, regardless of octane, that exceeds the following lead-content restrictions after the given date:
   2.0 gm/gal after 1 January 1974
   1.7 gm/gal after 1 January 1975
   1.5 gm/gal after 1 January 1976
   1.25 gm/gal after 1 January 1977
2. After 1 January 1974, no very low-octane (RON $\leq$ 91) gasoline may be sold with a lead content above 0.05 gm/gal or a phosphorus content above 0.01 gm/gal.
3. After 1 January 1974 retail outlets must, in general, be prepared to provide customers with a no-lead gasoline of at least 91 RON.[10]

Thus refiners must plan for the production of relatively low-octane, low-lead (or no-lead) gasolines, and this is the scenario adopted in choosing the major alternative specification for the model. On the other hand, in the studies done for and by the refining industry the tendency has been to consider only the alternative of maintaining present octane numbers while reducing or eliminating lead additions.[11] In addition, these studies have not gone into questions of residuals generation at all. Accordingly my discussion will fall into two parts. First, I attempt to assess the real-

[10] *Federal Register*, vol. 37, no. 36 (23 February 1972), p. 3883.

[11] One suspects that this approach was influenced by a desire to paint the industry's difficulties in the worst light in connection with the public battle with the auto companies over strategy. Thus the first major study in this area, "U.S. Motor Gasoline Economics," prepared by Bonner and Moore Associates for the API, June 1967, estimated an average cost of 2.3¢/gal to remove lead and maintain octane levels. This result may be contrasted with the costs of under 2¢/gal under almost all assumption sets estimated in J. C. Dunmyer, Jr., R. E. Froelich, and J. L. Putnam, "The Cost of Replacing Leaded Octanes" (paper presented at the 36th Midyear Meeting, API, Division of Refining, San Francisco, 13 May 1971).

ism of my results, as I did in chapter 5. To do this I compare my estimates of the cost of attaining two different lead-octane combinations, using the advanced technology refinery, with costs reported in two other studies.[12] I concentrate on the extremes currently being discussed: maintenance of present octane levels with zero lead additions and reduction to very low octanes also with zero lead. (I use 92 RON for "premium" and 90 RON for "regular" in this option; more likely there would be a single gasoline of somewhere between 90 and 92 RON.) But I also report results for the maintenance of present octanes with lead limits of 0.5 cc/gal and 0.25 cc/gal for an additional comparison with the Dunmyer study.

In order to arrive at cost estimates for the high-octane, no-lead case, I am forced to allow the refinery to charge more than the 150,000 bbl/day of crude that have been used as capacity.[13] This is, of course, because the volume of high-octane unleaded blending stocks is limited by the fractioning characteristics of the crudes (and by the limit of 3,000 bbl/day on purchased n-butane), even given the availability of hydrogen-intensive technology. (The results are presented more fully in table 24.)

Table 23 shows how my cost estimates compare with those of the other studies. The very close agreement between my results and those of the Dunmyer study *per gallon of gasoline* for the high-octane, no-lead case should be taken with a grain of salt because the underlying models differ on some points. In addition to the crude characteristics and new process availability noted above, the assumed TEL limits for present gasolines differ; they assume a 3 cc/gal constraint, while I have used 2.5 cc/gal. Nonetheless it is encouraging to see that for both extreme cases my cost results are in the proper neighborhood.

The next step is the comparison of residuals generation and other characteristics of my own solutions for the advanced technology with the following lead-octane requirement (referred to by the shorthand designations G1, G2, G3):

> G1: 100–94 octane; 2.5 cc lead upper limit
> G2: 92–90 octane; 0 cc lead upper limit
> G3: 100–94 octane; 0 cc lead upper limit

---

[12] Dunmyer, Froelich, and Putnam, "Cost of Replacing Leaded Octanes"; and W. L. Nelson, "Gasoline Deleading Remains Costly," *Oil and Gas Journal*, vol. 69, no. 18 (3 May 1971), pp. 118–23.

[13] This is consistent with Nelson, "Gasoline Deleading," although my increase in the crude charge for unleaded gasoline is considerably larger than the 7 percent he predicts. Dunmyer, Froelich, and Putnam do not have to increase the crude charge. This appears to be due to (1) the characteristics assumed for their single crude; and (2) their introduction of pentane (C_5) and hexane (C_6) isomerization to produce high-octane blending stocks from the straight-run gasoline fraction.

TABLE 23. Comparison of Cost Estimates for Reducing or Eliminating the Lead Content of Gasoline

(dollars)

| Cost | 40% 100-octane premium, 60% 94-octane regular | | | | | | | 40% 92-octane premium, 60% 90-octane regular | |
| | 0 cc TEL | | | 0.25 cc TEL | | 0.50 cc TEL | | 0 cc TEL | |
	Advanced refinery model	Dunmyer et al.[a]	Nelson[b]	Advanced refinery model	Dunmyer et al.[a]	Advanced refinery model	Dunmyer et al.[a]	Advanced refinery model	Nelson[b]
Per bbl crude	0.206	0.282	0.62	0.086	0.043	0.069	0	0.076[c]	0.08[c]
Per gal gasoline	0.0110	0.0114	0.029[d]	0.0046	0.0018	0.0033	0	0.0032[c]	0.0037[c]

[a] J. C. Dunmyer, Jr., R. E. Froelich, and J. L. Putnam, "The Cost of Replacing Leaded Octane" (paper presented at the 36th Midyear Meeting, API, Division of Refining, San Francisco, 13 May 1971). Based on extrapolation to 150,000 bbl/day size.
[b] W. L. Nelson, "Gasoline Deleading Remains Costly," Oil and Gas Journal, vol. 69, no. 18 (3 May 1971), pp. 118–23. No refinery size given.
[c] Cost decrease.
[d] Based on gasoline yield equal to half the crude volume.

All these requirements involve minimum production of 51,150 bbl/day of regular and 35,100 bbl/day of premium. The comparisons are summarized in table 24.

One of the major effects of the lead-octane constraints is on the refinery's energy use. Fresh heat is substituted for TEL as the more heat-intensive processes (especially catalytic reforming) are used to increase the volume of high clear-octane-number blending stocks. This substitution, in the absence of effluent charges or controls, is reflected directly in an increased discharge of residual heat and, through the purchase of high-sulfur residual fuel oil, in increased emissions of SO_2 and particulates.[14] Hydrocracking is adopted to help meet the low-octane, no-lead constraints, and becomes quite important in the high-octane, no-lead case. There is a corresponding decline in catalytic cracker use. The impacts of this shift may be seen in the increased sulfur recovery (more sulfur is removed from the oil and recovered from hydrocracker gases); in higher hydrogen sulfide discharges per barrel; and in lower phenols and BOD discharges per barrel (phenols tend to be broken down in the hydrogen-intensive processes). Even where residuals discharges per barrel decline from G2 to G3 (phenols, BOD, and particulates), the increase in the necessary quantity of crude charged results in increases in total daily discharges from the refinery. Of course, where per barrel amounts increase, the increased crude charge exacerbates the effect. This is an important point, since if the internal combustion engine retains its preeminent place in the transportation picture, and if gasoline demand (private auto travel demand) proves to be price inelastic, one implication of EPA's deleading schedule is a considerable increase in refinery discharges across the board—in the absence of emission charges or limits.

CHANGING QUANTITY AND QUALITY REQUIREMENTS
FOR OTHER PRODUCTS

So far only the implications of changing the quality constraints on gasoline have been discussed. Constant quantities of gasoline and other major products have been assumed, and no quality constraints have been put on those other products. In this section I explore the implications of

---

[14] In passing to G3 the increase in fresh heat inputs is more than compensated for by the decline in cat cracker (hence catalyst regeneration) activity, and particulates per barrel of crude decline.

TABLE 24.   Effects of Changing Lead and Octane Requirements of Gasoline

	Lead-octane requirements		
	Advanced G1	Advanced G2	Advanced G3
Cost of requirement/bbl crude ($/bbl)	0	($0.076)[a]	$0.206
Barrels of crude charged	150,000	150,000	191,700
Production			
Premium gasoline (bbl/day)	35,100	35,100	35,100
Regular gasoline (bbl/day)	51,150	51,100	51,150
Kerosene (bbl/day)	16,500	16,500	28,600
Distillate fuel oil (bbl/day)	15,000	15,000	15,000
Residual fuel oil (bbl/day)	3,000	3,000	3,000
Inputs			
Heat purchased (10^6 Btu/bbl)	0.231	0.292	0.340
Total fresh heat input (10^6 Btu/bbl)	0.399	0.445	0.487
Hydrogen manufactured (lb/bbl)	0	0	0.107
By-products			
Sulfur recovered (LT/day)	42.45	41.25	59.19
Sulfur recovered (LT/1,000 bbl)	0.283	0.275	0.309
Straight-run gasoline sold as petro- chemical feed (bbl/bbl)*	0.116	0.114	0.230
Process units			
Cat cracking, fresh feed (bbl/bbl)	0.461	0.369	0.295
Hydrocracking feed (bbl/bbl)	0	0.056	0.142
Coking feed (bbl/bbl)	0.124	0.133	0.138
H-oil feed (bbl/bbl)	0	0	0
Reformer feed (bbl/bbl)	0.139	0.193	0.243
Alkylation product (bbl/bbl)	0.080	0.083	0.079
Residuals			
Heat: (10^6 Btu)	46,650	50,400	66,200
(10^6/bbl)	0.311	0.336	0.346
Sulfur: (lb)	450	600	960
(lb/bbl)	0.003	0.004	0.005
Phenols: (lb)	5,100	5,250	5,940
(lb/bbl)	0.034	0.035	0.031
Oil: (lb)	7,350	7,800	10,300
(lb/bbl)	0.049	0.052	0.054
BOD: (lb)	9,450	9,600	11,100
(lb/bbl)	0.063	0.064	0.058
NH_3: (lb)	3,000	4,650	7,280
(lb/bbl)	0.020	0.031	0.038
SO_2: (lb)	214,800	226,500	308,000
(lb/bbl)	1.432	1.510	1.607
Particulates: (lb)	63,000	63,150	78,400
(lb/bbl)	0.420	0.421	0.409

Note: "Per bbl" refers to crude charged.

[a] Cost decrease.

* Low-octane gasoline-blending stocks may be sold as petrochemical feedstocks. This is consistent with Nelson, who notes that these blending stocks may be sold for a "2–4¢ per gallon lower price" (W. L. Nelson, "Gasoline Deleading Remains Costly," *Oil and Gas Journal*, vol. 69, no. 18 [3 May 1971], pp. 118, 120). It is assumed here that the price obtainable is 11.5¢ per gallon, which is 2.3¢ per gallon lower than the average for premium and regular gasoline.

a dramatic change in the refinery's product mix—involving both quantity and quality. In particular I assume a reduction in gasoline output to 60 percent of the benchmark figure (to 21,060 bbl/day of premium and 30,690 bbl/day of regular); increases in the production of both distillate and residual fuel oils; and a requirement that the sulfur content of distillate fuel oil be less than 0.3 percent and that of residual fuel oil less than 0.5 percent. (All gasoline is assumed to be no-lead, low-octane.)

Such a shift in product mix has been predicted, at least in qualitative terms, by petroleum industry authorities. For example, according to Joe F. Moore of Bonner and Moore consultants:

> It is quite possible that the average behavior of the domestic refining industry during the next 5 years will be to keep crude runs near present levels and to progressively increase distillate and residual fuel yields to meet the growing demand for these products while successively reducing gasoline yields as pool-octane requirements rise due to the demand for unleaded gasoline.[15]

Table 25 shows the implications of such a shift for the process unit mix, quantities of certain important inputs and by-products, and residuals discharges. The first column gives the data for the no-lead, low-octane solution from table 24 as a basis for comparison. The second column shows the effects of cutting gasoline production by 40 percent, increasing distillate fuel oil production by 50 percent (while requiring that its sulfur content be less than 0.3 percent), and doubling residual fuel oil production (requiring its sulfur content to be less than 0.5 percent). The biggest differences between the first two columns involve:

1. An increase in sulfur recovered.
2. An increase in the quantity of straight-run gasoline sold for petrochemical feed.
3. A decrease in total and purchased fresh heat input.
4. A switch away from the hydrocracking of gas oils.
5. A decline in the discharge of heat, phenol, oil, and particulates.

Smaller changes occur in the processing route for reduced crude and in the discharges of the other residuals.

The underlying reasons for these shifts may be summarized as follows:

1. The smaller required production of gasoline allows the sale of much more straight-run gasoline *and* a reduction in the necessary level of

---

[15] Joe F. Moore, "Deleading Can Have Antitrust Implications," *Oil and Gas Journal*, vol. 70, no. 32 (7 August 1972), p. 46.

operation of processes designed to produce very high-octane blending stocks. This in turn has a major effect on heat demand (and discharge) and on combustion residuals (SO_2 and particulates).

2. More reduced crude must be used to make residual fuel oil, leading to less coking. The lower availability of coke accentuates the decline in particulate emissions, but the increased burning of pitch (with higher particulate generation per barrel than residual fuel oil) works in the opposite direction.

3. Sulfur recovery and H_2S discharges to water increase because of the increased amount of hydrotreating necessary to meet the new quality constraints on the fuel oils.

Between columns 2 and 3 the increase in the required production of low-sulfur distillate fuel oil has very little effect, except insofar as the decline in catalytic cracking results in generally lower residuals discharges—especially heat, phenols, BOD, SO_2, and particulates. There is a substantial reduction in the sale of straight-run gasoline as the increasing requirements for distillate fuel oil make it necessary to meet the gasoline volume and octane constraints by blending more low-octane, straight-run stock with higher-octane products of the reformer and alkylation unit.

Column 4 shows the impact of adding to the third column's requirements a constraint that all kerosene be subject to hydrodesulfurization, which brings its sulfur content to below 0.1 percent and makes it suitable for sale as jet fuel. This increase in hydrotreating is the only processing change required to make the transition, and all the observed changes (in sulfur recovered, heat required and discharged, and other residuals discharges) are caused by this shift.

Finally, in column 5 the output of low sulfur residual fuel oil is increased from 6,000 bbl/day to 16,500 bbl/day. This causes extensive changes in required processing routes, and these in turn imply large changes in inputs and residuals discharges. To summarize the process changes first:

1. Total cracking of gas oil drops, and hydrocracking takes over almost entirely from catalytic cracking. (The object is to increase the production of naphtha for reformer feed.)

2. Coking drops by 50 percent and H-oil processing goes to zero as reduced crude is withdrawn for sale as residual fuel oil.

3. The reformer feed rate nearly doubles, and the alkylation product

TABLE 25. Effects of Changes in the Quality and Quantity of Product Mix

	Advanced G2[a] (1)	Advanced G2, low gasoline, higher fuel oil (2)	Advanced G2, low gasoline, high distillate fuel oil (3)	Advanced G2, low gasoline, high distillate fuel oil, low-S kerosene (4)	Advanced G2, low gasoline, high distillate fuel oil, high residual (5)
Cost per bbl of crude of requirements over cost of col. 2 ($/bbl)			0.527	0.567	1.045
Marginal cost of requirements ($/bbl)			0.527	0.040	0.478
Production					
92-octane premium (bbl)	35,100	21,060	21,060	21,060	21,060
90-octane regular (bbl)	51,150	30,690	30,690	30,690	30,690
Kerosene (bbl)	16,500	16,500	16,500	16,500[b]	16,500[b]
Distillate fuel oil (bbl)	15,000	22,500[b]	40,500[b]	40,500[b]	40,500[b]
Residual fuel oil (bbl)	3,000	6,000[b]	6,000[b]	6,000[b]	16,500[b]
By-products					
Sulfur recovered (LT)	41.25	72.00	62.52	66.72	79.83
Straight-run gasoline sold as petrochemical feed (bbl/bbl)	0.114	0.240	0.170	0.170	0.140
Inputs					
Heat purchased (10⁶ Btu/bbl)	0.292	0.224	0.224	0.242	0.447
Total fresh heat (10⁶ Btu/bbl)	0.445	0.350	0.349	0.367	0.501
Hydrogen manufactured (lb/bbl)	0	0	0	0	0.799

Process Units					
Cat cracking fresh feed (bbl/bbl)	0.369	0.441	0.263	0.262	0.004
Hydrocracking feed (bbl/bbl)	0.056	0	0	0	0.129
Coking feed (bbl/bbl)	0.133	0.106	0.106	0.105	0.051
H-oil feed (bbl/bbl)	0	0.009	0.009	0.010	0
Reformer feed (bbl/bbl)	0.193	0.135	0.135	0.135	0.253
Alkylation product (bbl/bbl)	0.083	0.005	0.032	0.032	0.048
Hydrotreating feed (bbl/bbl)	0.139	0.219	0.206	0.316	0.370
Heat (10^6 Btu/bbl)	0.336	0.234	0.221	0.230	0.260
Sulfide (lb/bbl)	0.004	0.005	0.005	0.005	0.008
Phenol (lb/bbl)	0.035	0.024	0.018	0.018	0.009
Oil (lb/bbl)	0.052	0.037	0.035	0.036	0.040
BOD (lb/bbl)	0.064	0.075	0.063	0.064	0.064
NH_3 (lb/bbl)	0.031	0.037	0.034	0.036	0.057
SO_2 (lb/bbl)	1.510	1.373	1.168	1.226	1.394
Particulates (lb/bbl)	0.421	0.346	0.303	0.303	0.134

Note: "Per bbl" refers to crude charges.
[a] From table 24.
[b] Low sulfur.

volume goes up by 50 percent as greater quantities of very high-octane blending stocks are required to meet the gasoline constraints. (Gasoline volume is maintained by blending these stocks with more of the straight-run gasoline previously sold as petrochemical feed.)

4. Hydrotreating increases because of the greater volume of reduced crude requiring desulfurization.

These changes imply a very large increase in required fresh heat input, a shift to the in-plant production of hydrogen (the quantity recovered from the reformer gases being no longer sufficient to meet all process demands), and an increase in sulfur recovery from hydrotreater off-gases. The impact on residuals is particularly interesting, since both increases and decreases in discharges per barrel of crude are observed. Heat, sulfide, and NH_3 discharges go up, primarily because of the heavier reliance on hydrogen- and heat-intensive processes. (Recall that the hydrogen processes, in removing sulfur from the oil, produce gas streams high in H_2S, and that some of this gets dissolved in condensed process streams and becomes a waterborne residual problem. I have assumed that NH_3 concentration is proportional to H_2S concentration.) SO_2 and oil discharges increase because of increased heat input and discharge respectively. Phenols discharges decrease because it is assumed that the hydrogen-intensive processes reduce phenols formation, and hence their condensates are low in phenols relative to the coker and cat cracker. Particulates also decrease, and this is simply the result of the substitution of purchased residual fuel oil for refinery coke as a source of fresh heat.

In the next two chapters, when I examine the impact of discharge limits and emission charges on the refinery's residual discharges, I use the advanced refinery and the product mix from column 5 in table 25 as one of the two cases to be examined. The other will be the basic refinery with the benchmark product mix.

CHANGING QUALITY OF INPUTS: THE SULFUR CONTENT
OF CRUDE OIL

Since crude oil is quantitatively by far the most important input to the refinery, it is reasonable to expect that changes in its quality will have a significant effect (ceteris paribus) on the refinery's residuals discharge. In this section I explore the effects of changing the sulfur content of the crude charged and find this expectation fulfilled.[16] I do not explore any of the other quality differences in crude oils that could affect residuals; for example, nitrogen, heavy metals, or brine content. The method used is simply to take advantage of the price differential between the crudes originally built into the model and to expand the "available" quantity of the cheaper, higher-sulfur crude in steps. This approach is adequate for the present concern but, as discussed in chapter 8, it is necessary to explore the effects of the relative prices more carefully when one is interested in affecting the sulfur residuals directly by effluent charges or discharge limits.[17]

I report here the results for varying crude availability in the basic refinery under the benchmark set of product constraints. This suffices to give the flavor of the changes.

Table 26 sets out the most important changes resulting from increasing the availability of the Arabian mix crude to 43,326 bbl/day—the maximum that can be charged while still meeting the constraints—with one intermediate step at 30,000 bbl/day. The following effects are worth noting:

1. Costs: Because of the very large price differential, costs decrease significantly with availability of the higher-sulfur crude. The change in costs attributable to each additional barrel of high-sulfur crude is about $0.55 at zero availability and falls to $0.54 when 30,000 bbl/day are readily available. Each barrel lowers costs somewhat less than the price differential ($0.73/bbl) because the lower quality and different fractioning characteristics of the higher-sulfur crude mean that other inputs must be increased—processing must be more heat intensive and TEL additions must increase—in order to meet the other constraints. Similarly, because it gets progressively tougher to meet these constraints as

---

[16] As noted, I do not maintain all other conditions exactly equal, for the crudes differ slightly in fractioning characteristics as well as in sulfur content.

[17] Clearly, as the price differential shrinks, the refinery will find it profitable to switch to the lower-sulfur crude to cut SO_2 (or sulfide) emissions at lower emission charge levels. Indeed at some price the higher-sulfur crude will not be used at all, even in the absence of SO_2 emission charges or constraints, because of the link between sulfur content and quality of gasoline-blending stocks. For the basic refinery this price is $3.57/bbl of the higher-sulfur crude. In other words the refiner prefers the higher-quality crude when there is still a differential of about $0.18/bbl. According to W. L. Nelson, "Sulfur Penalties for Average Crudes," *Oil and Gas Journal*, vol. 69, no. 36 (6 September 1971), p. 108, the average price for sulfur content (in elemental and mercaptan forms) would be 11.3¢/bbl lower for a crude with total sulfur content of 1.5 percent than for one with a total content of 0.4 percent.

TABLE 26.   Effects of Changing the Availability of Cheaper, Higher-Sulfur Crude Oil: Basic Refinery, Benchmark Product Mix

	Upper limit on availability of high-sulfur crude (*bbl/day*)		
	0	30,000	43,326
Average change in costs (*$/bbl*)	—	0.109	0.157
Marginal change in costs (*$/bbl*)	0.549	0.540	0
Average TEL content of gasoline (*cc/gal*)	2.31	2.43	2.50
Kerosene produced (*bbl/bbl*)	0.1000	0.1040	0.1058
Sulfur recovered (*LT*)	24.42	35.73	42.06
Purchased heat (*10^6 Btu/bbl*)	0.223	0.226	0.226
Total heat (*10^6 Btu/bbl*)	0.386	0.390	0.413
Reforming charge (*bbl/bbl*)	0.133	0.137	0.139
Alkylate production (*bbl/bbl*)	0.074	0.076	0.076+
Residuals			
Heat (*10^6 Btu/bbl*)	0.296	0.283	0.287
Sulfide (*lb/bbl*)	0.002	0.002	0.003
Phenol (*lb/bbl*)	0.031	0.032	0.032
Oil (*lb/bbl*)	0.046	0.047	0.047
BOD (*lb/bbl*)	0.058	0.060	0.061
NH$_3$ (*lb/bbl*)	0.012	0.018	0.022
SO$_2$ (*lb/bbl*)	1.147	1.345	1.469
Particulates (*lb/bbl*)	0.415	0.420	0.424

the daily charge contains a higher proportion of the lower-quality crude, the marginal profit addition falls with availability. Since the level of use of the second crude cannot go above about 43,326 bbl/day, the marginal addition to profit due to an increase in availability above that level is zero.

2. Production: As the use of higher-sulfur crude increases, average TEL content rises (by 8 percent between zero barrels per day and 43,330 bbl/day use). Indeed the refinery can charge no more of the higher-sulfur crude than this because the constraints on product quality, in particular gasoline lead and octane requirements, cannot be met with higher charges.[18] Kerosene production increases as, of course, does the recovery of sulfur. This latter change reflects directly the increasing input of sulfur

---

[18] The advanced refinery required to make low-octane, unleaded gasoline plus 16,500 bbl/day of kerosene, 15,000 bbl/day of distillate fuel oil, and 3,000 bbl/day of residual fuel oil has even less tolerance for the higher-sulfur crude.

to the refinery and the increase in recovery is about 42 percent over the full range of crude charges.[19]

3. Processing: Only minor changes occur in processing and these are attributable to differences in the relative importance of the several fractions between the two crudes. For example, the use of the catalytic reformer increases by only 4.5 percent between a charge of zero barrels per day and 43,330 bbl/day of higher-sulfur crude, because the volume percentage of naphtha in that crude is higher than in the lower-sulfur oil.[20]

4. Inputs other than crude oil: Total fresh heat applied per barrel of crude increases by 6.5 percent over the charge range. As noted, this may be attributed to the lower quality and different fractioning characteristics of the higher-sulfur crude. Thus the increase in reforming results in greater heat use and arises from the larger naphtha volume per barrel of this oil. More alkylation also means greater heat application and may be attributed to the need to enhance the gasoline-blending possibilities by producing more of this very high-octane stock.

5. Residuals: The most important change here, of course, is in SO_2 emissions caused by the burning of sour coke from the coker and the higher-sulfur coke deposited on cat cracker catalysts. Sulfide discharged in waste condensate also increases, as one would expect. Phenol and BOD discharges go up at the same time, largely because I assumed a rough correlation between sulfide, phenol, and BOD concentrations in several sour-water streams. The oil residual goes up because of the increased cooling requirements. (Recall that a small average oil content is assumed for cooling water.)

CHANGING INPUT COSTS: PURCHASED FRESH HEAT

As the reader has no doubt realized by now, heat is also a very important input to the refining process. Changes in heat requirements tend to reflect changes in the difficulty of meeting product constraints; and changes in the mix of purchased and in-house heat supplies occur as product mix, technology, and crude quality vary. In this section the results of

[19] This is less than the percentage increase in sulfur input because of changes in the fate of the sulfur due to differences in the fractioning characteristics of the two crudes. Proportionately more of the sulfur in the high-sulfur crude leaves the refinery in the products.

[20] The effect here of the increasing volume of naphtha is partially offset by a decline in the reduced crude volume and hence in coker gasoline, which is also charged to the reformer (after hydrotreating).

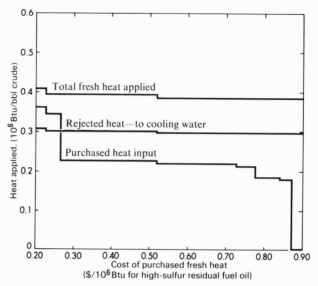

Figure 14. Response to changes in the cost of purchased heat: basic refinery, benchmark product mix.

operating directly on the heat input by varying the cost of purchased heat are considered.

In the benchmark situation it was assumed that the refiner is able to buy residual fuel oil in three different sulfur grades: a 2 percent grade for $0.477/10^6 Btu, a 1 percent grade for $0.593/10^6 Btu, and a 0.5 percent grade for $0.661/10^6 Btu. But in the southwestern oil fields, where natural gas may be purchased cheaply, the cost per 10^6 Btu might be much lower, perhaps as low as $0.20–$0.25. It will be of considerable interest, then, to observe what happens as the assumed cost for the higher-sulfur fuel oil is raised from $0.20 to $0.90/10^6 Btu (at which point heat purchases have gone to zero).[21] I report the results in detail for the basic refinery, with the benchmark set of product specifications.

Two types of response are used by the refiner in adjusting to heat cost increases: the substitution of in-house heat sources for purchased heat; and the substitution of capital, in the form of larger heat exchangers and more efficient furnaces, for fresh heat inputs. In figure 14 and table 27, the specific responses, their timing, and effects are detailed.

The figure and table are largely self-explanatory; it need only be emphasized that the quantity of *purchased* heat is very responsive to price

[21] In order to avoid a spurious switch to a low-sulfur residual, I maintain the same price ratio between the grades during the incremental process.

TABLE 27. Response to the Increasing Cost of Fresh Heat: Process Changes
and Other Adjustments

Cost of higher-sulfur purchased heat source ($/10^6 Btu)	Action taken by refiner
0.228	Chooses higher-capital fractioning process
0.266	Burns sweet coke rather than sell
0.519	Chooses higher-capital reforming process
0.726	Burns some high-sulfur reduced crude instead of coking
0.775	Burns polymer rather than sell
0.841	Burns some refinery gas before desulfurization
0.872	Burns enough refinery gas (after desulfurization) to decrease purchased heat to zero

changes of the magnitude assumed here, while total heat applied changes by a *total* of only 5.8 percent over the entire range. If rough arc elasticities are calculated over this price range for both total and purchased heat, the former is about -0.05 and the latter -1.57, or over 30 times as large.[22]

Table 28 summarizes a few other effects of the refiner's response to the increasing cost of purchased fresh heat. Costs, of course, increase, the magnitude of the effect being between 0.35¢/bbl per penny increase in cost when the heat cost is low and 0.08¢/bbl per penny increase in cost between $0.85 and $0.90/10^6 Btu. There is a slight decline in BOD at a heat cost of $0.726/10^6 Btu due to the shift in processing mix required when some reduced crude is burned. SO_2 emissions decline slowly but consistently as heat requirements go down and as in-house fuels, lower in sulfur per 10^6 Btu are substituted for the purchased residual with 2 percent sulfur.

Particulate emissions from the basic refinery increase dramatically when sweet coke is substituted for the purchased residual. They ultimately decrease slightly as heat requirements are reduced and as lower-particulate in-house fuels are substituted. There is a small decline between $0.70/10^6 Btu and $0.75/10^6 Btu due to a small shift toward lower recycle cat cracking.

CHANGING INPUT COSTS: WATER WITHDRAWALS

There is a long tradition of interest among natural resource researchers in the *volume* of water withdrawals (and discharges) attributable to vari-

---

[22] Because of the step-wise character of the two heat "demand curves," it is not particularly useful to calculate elasticities over short segments.

TABLE 28.   Cost and Residuals Discharge Changes Due to Increases in the
   Cost of Fresh Heat

Cost of high-sulfur fresh heat purchase ($/10⁶ Btu)	Total cost increase ($/bbl crude)	BOD discharge (lb/bbl crude)	SO₂ discharge (lb/bbl crude)	Particulate discharge (lb/bbl crude)	Heat discharge to water (10⁶ Btu/bbl crude)
0.20	0	0.0603	1.476	0.196	0.306
0.25	0.018	0.0603	1.441	0.195	0.300
0.30	0.031	0.0603	1.429	0.423	0.300
0.35	0.042	0.0603	1.429	0.423	0.300
0.40	0.054	0.0603	1.429	0.423	0.300
0.45	0.065	0.0603	1.429	0.423	0.300
0.50	0.076	0.0603	1.429	0.423	0.300
0.55	0.087	0.0603	1.415	0.423	0.297
0.60	0.098	0.0603	1.415	0.423	0.297
0.65	0.109	0.0603	1.415	0.423	0.297
0.70	0.120	0.0603	1.415	0.423	0.297
0.75	0.131	0.0601	1.415	0.420	0.297
0.80	0.141	0.0601	1.415	0.420	0.297
0.85	0.150	0.0600	1.390	0.418	0.297
0.90	0.154	0.0600	1.010	0.408	0.297

Note: Basic technology, benchmark product mix.

ous activities. In the case of industry the problem of waterborne residuals discharges is probably far more significant than that of withdrawal volume; nonetheless it will be appropriate to consider in this chapter the effect on the refinery of increases in the costs of water withdrawals. In order to do so very simply I vary the costs of all three of the separate withdrawal streams (cooling water, desalter water, and boiling water) at once. I begin with the price set used in most of the computations:

Cooling water	$0.015/1,000 gal
Desalter water	$0.070/1,000 gal
Boiler water (for steam)	$0.15/1,000 gal

The cost figures are assumed to include the cost of pre-use treatment, and this produces the differential between the three streams since cooling water requires virtually no treatment for once-through use,[23] while water for the boiler must be subject to some demineralization to cut down on tube scaling and so forth. These costs are increased in steps of $0.01, $0.05, and $0.10/100 gallons respectively up to $0.095, $0.470 and $0.95/1,000 gallons respectively. These increases can be thought of as rep-

[23] See, for example, R. M. Silverstein and S. D. Curtis, "Cooling Water," in *Chemical Engineering*, 9 August 1971, pp. 84–94. On p. 88, treatment costs for a once-through system of 72,000 MGD are estimated to be between 0.035¢ and 0.105¢/1,000 gal.

resenting increasing treatment costs occasioned by declining water quality in the withdrawal source or increasing "abstraction charges" levied
by some regional water management authority.[24] Whether the particular
increases are realistic is beside the point since the interest here is in tracing the response of the refinery. One can always visualize a situation in
which a drought emergency makes it necessary for a refiner to reduce his
withdrawals, and in such a situation the shadow price attached to the
withdrawal constraint would rise and would be roughly analogous to the
increasing cost figures.[25]

The effects of these cost changes are shown in table 29 (for the basic
refinery with the benchmark product mix) where I indicate both the
water use pattern at five different cost sets and the costs at which significant adjustments in that pattern are made. The changes run as follows: First, at a cost of $0.039/1,000 gal for cooling water, installation of
a cooling tower becomes profitable, cutting withdrawals dramatically.[26]
Next, at $0.040/1,000 gal (the assumed cost of this recirculation alterna-
tive), withdrawals are further reduced by the recirculation of effluent
from the API-separator and sour-water stripper treatment steps to cooling tower makeup. Then, at much higher cost sets, cooling tower makeup
withdrawals increase as the API separator effluent is, instead, subject to
secondary and tertiary treatment and recirculated to the boilers. These
changes go on up to a cost of $0.088/1,000 gal for cooling water (a boiler
water cost of $0.88/1,000 gal) after which all possible streams are being
handled in that way. No further changes occur up to the $0.095, $0.470,
$0.95 price set, although certainly if the costs went higher options would
still be open to the refiner. For example, he could choose to install the
higher-capital coker to take advantage of its lower heat rejection rate.

[24] See Lyle Craine, *Water Management Innovations in England* (RFF, 1969), for a
discussion of an abstraction fee system being introduced in England.
[25] But not exactly analogous since I do not separate the costs of treatment and withdrawal per se. To the extent that only *quantity* were affected by a drought, only the
withdrawal costs would be increased and the relative effects on the three streams
would be quite different. Boiler water withdrawals would become relatively cheaper.
[26] This result depends, of course, on my estimate of the cost of recirculation in cooling towers; but I believe my estimate to be conservatively high since it includes the
cost of treating tower blowdown in an API separator.
When cooling towers are operated the evaporation required to cool the water to (or
nearly to) ambient wet bulb temperature takes place within the plant, and thus "consumptive use" of water appears to rise. But if heated water is discharged to a water
course in the same area, a similar amount of evaporation will take place. (The two
amounts generally will not be the same because of such complications as bank cooling
in the stream, etc.) Thus it is misleading to state without qualification that an "increase" in consumptive use occurs when cooling towers are introduced.

TABLE 29. Refinery Response to Increases in the Cost of Water Withdrawals: Basic Refinery, Benchmark Product Mix

	Cost of water ($/1,000 gal)					
	Cooling, 0.015; desalter, 0.07; boiler, 0.15	Cooling, 0.045; desalter, 0.22; boiler, 0.45	Cooling, 0.055; desalter, 0.27; boiler, 0.55	Cooling, 0.065; desalter, 0.32; boiler, 0.65	Cooling, 0.075; desalter, 0.37; boiler, 0.75	Cooling, 0.095; desalter, 0.47; boiler, 0.95
Cost/bbl of crude ($)	— *a	0.0309 *b	0.0336 *c	0.0355 *d	0.0367 *e	0.0392 *e
Cost/bbl/$0.01 increase in cooling water cost	—	0.010	0.0084	0.0071	0.0061	0.0049
Capital intensity						
Fractioning tower	High	High	High	High	High	High
Reformer	Low	Low	Low	High	High	High
Coker	Low	Low	Low	Low	Low	Low
Purchased heat (10^6 Btu/bbl)	0.226	0.224	0.224	0.219	0.219	0.219
Water volume (1,000-gal units)						
Cooling water withdrawals	154,350	12,320	12,680	14,070	14,130	14,130
Recirc. to cooling tower from API separator	—	1,960	1,600	60	10	0
Desalter withdrawals	360	360	—	120	120	120
Recirc. to desalter from secondary treatment	—	—	360	—	—	—
Boiler feedwater withdrawals	2,661	2,661	2,661	303	243	237
Recirc. to boiler from carbon adsorption	—	—	—	2,360	2,520	2,530
Residuals discharged						
Heat (10^6 Btu/bbl)	0.300	0	0	0	0	0
Sulfide (lb/bbl)	0.003	neg.	neg.	neg.	neg.	neg.
Phenol (lb/bbl)	0.032	0.005	0.005	0.005	0.005	0.005
Oil (lb/bbl)	0.047	0.011	0.011	0.009	0.009	0.009
BOD (lb/bbl)	0.060	0.032	0.031	0.010	0.008	0.008
NH$_3$ (lb/bbl)	0.021	0.014	0.012	0.002	neg.	0
SO$_2$ (lb/bbl)	1.429	1.429	1.429	1.415	1.415	1.415
Particulates (lb/bbl)	0.423	0.423	0.423	0.428	0.428	0.428

Note: The pattern is substantially the same for other combinations of assumptions about refining technology and product mix. In particular, the costs at which changes occur are exactly the same, since the water-use technology is not assumed to change between cases. The quantities of water and residuals involved at any particular water-cost level will, of course, differ across the cases.

* In summarizing the costs at which shifts occur, I use the cooling water withdrawal cost to stand for the entire vector. Thus the first shift occurs at the point of $0.039, $0.19, $0.39 costs/1,000 gal for cooling, desalting, and makeup water respectively.

[a] To cooling tower @ $0.039; to recirc. from API separator to cooling tower makeup @ $0.040.

[b] To secondary treatment + recirc. to desalter—$0.048–$0.051.

[c] To secondary + tertiary treatment + recirc. to boiler—$0.0562–$0.0637; desalter withdrawals non-zero after $0.0619; to high capital reformer @ $0.0556.

[d] Additional streams treated to tertiary level and recirculated to boiler feed—$0.0704.

[e] Additional streams treated to tertiary level and recirculated to boiler feed—$0.0875.

(Notice that the fractioning tower is the high-capital version at the base price set because of the cost of heat. The high-capital reformer is chosen at a cost of $0.056/$10^3$ gal for cooling water.)

The effects on waterborne residuals are also noted in the table. Basically, heat rejection *to the water* is completely eliminated, and significant reductions in the quantities of the other four residuals are obtained as well. These discharge reductions are produced both by recirculation to the cooling tower makeup and by treatment. In the former case it is assumed that the reduction results from the oxidation of hydrogen sulfide and the evaporation or oxidation of phenols in the course of close air-water contact in the cooling tower. The effects of secondary and tertiary treatment are, of course, straightforward. In this connection, the use of secondary treatment introduces *sludge*, and there is a small increase in particulate emissions because of sludge incineration. Incinerator residue accounts for the increase in solid residuals. There is only a minor effect on SO_2 that is entirely due to the adoption of high-capital reforming. If gaseous hydrocarbon emissions were traced, an increase due to cooling tower adoption would be found. In addition, the small amount of SO_2 implied by the oxidation of H_2S in the tower is ignored here.

A Postscript: The Impact of the Price of Recovered Sulfur

Under the assumptions about the sulfur recovery process discussed in chapter 4, the refiner finds it profitable to recover sulfur if the price he can obtain per long ton (2,240 lb) is greater than $5.88.[27] The difference in SO_2 emissions made by sulfur recovery in the absence of any direct pressures, such as constraints or effluent charges, amounts to about 179,000 lb/day for the basic refinery (with benchmark product mix). That is, emissions from the basic refinery would be 2.62 lb/bbl of crude instead of 1.43 lb if the sulfur price were below $5.88/long ton. Clearly, this is an important parameter in terms of its environmental quality implications.

The sulfur price at which recovery becomes profitable is quite sensitive to my assumption about the cost of the process (net of cooling, steam, and heat costs). Doubling the unit cost from $0.22 for 75.6 lb of sulfur ($6.50/long ton) to $0.44 for 75.6 lb of sulfur ($13/long ton) results in an

[27] This result depends on the profitability of gas desulfurization and on the requirement that sour-water stripping be carried out.

increase in the critical price to \$12.40 (or 250 percent). But since the current market price of sulfur is, as noted, about \$20/long ton, recovery would be profitable even under the extreme assumption that process costs have been underestimated by 50 percent.[28]

Now that I have explored the implications for residuals generation of changes in several important indirect influences, I am ready to turn to a consideration of the more familiar direct influences. It has been my aim in this chapter to support the first of my twin contentions: that many influences other than explicit environmental quality management actions will affect residuals generation by industrial plants. In the next two chapters, in the process of observing the refiner's response to emission charges or limits, I explore the second contention: that industrial response to explicit environmental quality actions will generally involve changes in processes or input qualities, the recirculation of residuals-bearing streams, or the production of by-products in addition to end-of-pipe treatment.

---

[28] The observation that sulfur recovery is probably profitable for refiners under current technology and the current sulfur market price is supported by evidence from the NAPCA inventory of air pollution sources in the Delaware River estuary region. Of the seven large refineries located along the estuary, four are reported to have sulfur recovery units, and the report for one of the three remaining refineries is obviously incomplete so that one cannot rule out the possibility that it also has such a unit.

VII

EFFECTS OF DIRECT INFLUENCES ON WATERBORNE RESIDUALS DISCHARGES

As pointed out originally, one of the principal aims of developing models of residuals generation and discharge by industrial plants has been to provide tools for use by regional environmental quality management authorities. Chapter 1 included a description of the framework within which these models were envisioned being used. The key role of the models is to show how discharges to the regional environment vary with changes in certain policy instruments: either effluent charges or discharge constraints (allocations). In line with the terminology adopted for discussing the petroleum refinery model in isolation, these policy instruments will be referred to as *direct* influences on residuals generation and discharge to distinguish them from such factors as input cost and output mix which were explored in chapter 6.[1] This chapter is the beginning of an investigation of the effects of direct influences and thus in a sense brings us to the heart of the model.

Given the complexity of the refinery and the number of residuals involved, I can hardly aspire to completeness in my investigation of direct influences. That is, I cannot hope to show how the refiner would respond to all possible combinations of charges or constraints on all residuals under all combinations of assumptions about the indirect influences. The

[1] The refiner's response to the imposition of an effluent charge may, of course, be to adjust his production processes, input quality, or output mix. Such a response, then, represents the other side of the coin examined in the last chapter; the indirect influences that were discussed as causes here become effects. The direct-indirect distinction is drawn on the basis of whether the policy is aimed at the refiner's discharges or at some other concern.

132

goal here will be the more modest one of illustrating the kinds of effects and the degrees of sensitivity observed. I concentrate on varying a single effluent charge (or discharge constraint) at a time, holding everything else fixed—and, most frequently, holding other effluent charges at zero. In addition, however, I show the results of some simultaneous variations in more than one direct influence and of changes in assumptions about the values of the indirect influences. Finally, because some potential users of these models may be interested in questions concerning industries rather than regions and thus may wish to consider the use of the models in isolation rather than in the larger framework, I illustrate as I go along some of the different questions one can ask the model. That is, I illustrate variations on the simple theme: Given effluent charge e on residual r, what is the vector of discharges from the refinery? Examples of such variations are: How much does it cost to reduce BOD discharges by x percent? Or all discharges by y percent?

Before proceeding to a discussion of the results, however, it will be valuable to deal with two subsidiary issues, one major and one minor. The first is the question of effluent charges versus discharge limits, and the second is the choice of a technique for displaying the results of the investigations.

There is no need to become involved in the continuing debate about the relative efficacy of effluent charges;[2] it will not aid in exposition and would only tend to obscure the fundamental issue of the potential utility, or lack of it, of the model. In what follows, however, it is necessary to use one device or the other for investigating the impact of direct influences on residuals discharges. The number of times each is used should not be interpreted as an index of my preference.

It is also important to note that from the point of view of the model the two instruments are completely symmetric.[3] This is simply an example of the general symmetry between constraints and objective function entries in any optimization problem, a principle that has been explained

[2] The best known proponent of effluent charges is Allen Kneese; see Allen V. Kneese and Blair T. Bower, *Managing Water Quality: Economics, Technology, Institutions* (John Hopkins Press for RFF, 1968). Much of the debate over standards versus charges is summarized in A. Myrick Freeman III and Robert H. Haveman, "Water Pollution Control, River Basin Authorities and Economic Incentives: Some Current Policy Issues," *Public Policy*, vol. 19, no. 1 (Winter 1971), pp. 53–74; and Marc J. Roberts, "Organizing Water Pollution Control: The Scope and Structure of River Basin Authorities, ibid., pp. 75–141.

[3] This statement must be qualified because of the discontinuities in the first derivatives implied by linearity. I do this below.

perhaps most clearly by Marglin.[4] This principle says that if a decision variable x in an optimization problem has the value $\hat{x}$ at the optimum associated with the price set including $\hat{p}_x(> 0)$, a constraint requiring $x \geq \hat{x}$ (for the same price set except $p_x = 0$) will have associated with it a shadow price $\hat{q}_x$ that will be equal to $\hat{p}_x$.[5] If $\hat{p}_x < 0$, the appropriate constraint (with $p_x = 0$) would be of the form $x \leq \hat{x}$, and so forth. Heuristically, this simply says that if in one situation there is a credit of $\hat{p}_x$ for producing x and the best decision is to produce $\hat{x}$ units, it would be possible in a situation in which the credit for producing x was zero but in which there was a requirement to produce $\hat{x}$ units to add an amount equal to $\hat{p}_x\Delta x$ to the objective function by *reducing* the required production level by some small amount Δx.

In a linear model this equivalence does not hold at every point because the first partial derivatives are discontinuous. Thus a given production level $\hat{x}$ will be chosen for a range of prices $\underline{p}_x$ to $\overline{p}_x$ and a particular shadow price q_x will be associated with a range of constraint levels $\overline{x}$ to $\underline{x}$. In particular, to understand this relation for effluent charge-discharge constraint pairs, consider the two graphs in figure 15.

Part A shows the relation between charges and constraints on a residual discharge, assuming an optimization model involving nonlinear functions with continuous first partial derivatives. Whether one specifies a charge and finds the implied discharge or specifies a discharge limitation and finds the implied shadow charge makes no difference.

Part B, on the other hand, shows the situation for a linear model of the type developed here. The first point to note is that if charges are specified only that portion of the curve shown by dotted lines is observed, whereas if constraints are specified only the solid portions are observed. Thus beginning with a zero charge and increasing its absolute value by small increments (making it larger negatively), no change in discharges from D_0 is observed until the charge reaches c_1, where there is a sudden fall in discharges to D_1. Similarly, as the charge is increased above c_1, the discharge is observed until c_2 is reached, at which level the discharge drops to D_2. But if one begins with a limit on discharges of D_0 and gradually decreases it, no shadow charge between zero and c_1 is observed, only c_1 itself, which is associated with the first small reduction below D_0. As the

[4] Stephen A. Marglin, *Public Investment Criteria* (Massachusetts Institute of Technology Press, 1967), chap. 1.
[5] If it is already known that $x = \hat{x}$ at the optimum for the given price set, the possibility that the constraint will not be binding need not be considered.

A. Nonlinear model, continuous partial derivatives

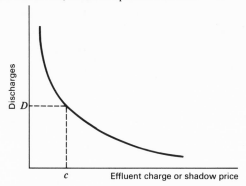

B. Linear model

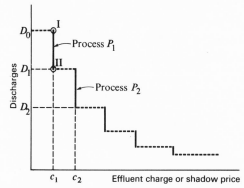

Figure 15. Effluent charges, discharge constraints, and shadow prices.

allowed discharges are reduced from D_0 to D_1, the shadow charge continues to be equal to c_1, but as soon as the allowed discharge falls to D_1, the shadow charge increases (in absolute value) to c_2. These discontinuities are associated with shifts between alternative processes within the linear model. In this example it has been assumed that there is some process, P_1, by which discharges may be reduced from D_0 to D_1, and that the *total* cost to the refiner of using P_1 is c_1 per unit of the residual removed.[6] Similarly, a process P_2 exists and is capable of reducing discharges by $D_1 - D_2$ units at a total cost of c_2 per unit removed.

Strictly speaking, there is no exact effluent charge–shadow price equivalence in the linear case because of discontinuities at the points of process

---

[6] The total is emphasized here because more than simply the unit cost of the process itself may be involved. Thus, for example, if P_1 produces a secondary residual with an effluent or disposal charge, the total cost of operating P_1 includes the cost of dealing with this new residual.

changeover. Effluent charges and discharge constraints, however, are still usefully considered to be symmetric in the terms given by the following statements:

1. Ceteris paribus, any level of discharge of a residual that is feasible may be obtained by using a discharge constraint. The shadow cost of that constraint will in general be the lower bound of possible marginal costs (per unit change in discharge) of attaining that level. For any particular problem and residual there will exist a finite set of discharge levels at which the shadow cost will be undefined, the marginal cost being a different value depending on whether the change in the constraint is assumed to be an increase or decrease.

2. Any value of an effluent charge may be attached to a particular discharge activity. The charge will, in general, be greater than or equal to the lowest charge that will produce a particular discharge level; and for a particular problem and residual, only a finite number of discharge levels will be obtainable via charges. For certain charge levels the appropriate discharge level will be undefined; i.e., when the charge exactly equals the cost of the available discharge-reduction process, the program will be indifferent among an infinite number of possible discharge levels.

3. It is possible to move arbitrarily close to a symmetric pairing of effluent charge and shadow cost by approaching the "corners" of figure 15, part B, from the proper direction. Thus for point I, let the effluent charge become larger (negatively) keeping it less in absolute value than c_1, hence maintaining discharge at D_0. On the other hand, increase allowed discharges toward D_0, thus maintaining the shadow cost at c_1. The shadow cost will remain greater than the effluent charge. Similarly, point II may be approached by decreasing the absolute value of the effluent charge toward c_1 and by decreasing the allowed discharge toward D_1 from above. As point II is thus approached the effluent charge will always be greater than the shadow cost.

One relatively minor question concerning possible methods of presenting results should be dealt with before turning to these results themselves. Because this model was originally formulated for use as part of a larger model of the regional residuals management problem, it is logical that the major concern should be with the absolute size of discharges for any combination of direct actions and assumptions about indirect influences. On the other hand, presentation of results in this form produces one difficulty in this book, for there may be interest here in comparing the sensitivities of two different residuals to similarly changing levels of effluent charges. For these purposes, percentage reductions in discharge are more valuable. If there is interest, however, in comparing the results of applying similarly varying effluent charges to the *same* residual under *different* sets of assumptions about the indirect influences (hence with different discharges in the absence of charges or controls), the percentage reduction approach will not provide sufficient information but should at least be supplemented by data on absolute discharge levels. In what follows, there is an attempt to mix these two techniques in such a way as to inform without overwhelming.

HEAT REJECTED TO COOLING WATER

There are two basic alternatives for reducing the discharge of heated cooling water from the refinery. One involves process changes designed to decrease the waste heat generated in processing (mainly through substitution of capital for fresh and residual heat inputs). The other involves the installation of cooling towers in which the cooling water gives up its heat load to the atmosphere, permitting subsequent recirculation back to the coolers. My results indicate that for a realistic range of cooling tower costs, only the second alternative is chosen in response to an effluent charge on heat discharges to a natural water body; cooling towers are simply very cheap relative to process changes involving more efficient heat use.

Since there are no differences in cooling tower unit costs caused by refinery process technology or product mix, it is sufficient to concentrate on the basic refinery with benchmark product mix. In that situation cooling water withdrawals are assumed to cost $0.015/1,000 gal and the basic cost of recirculation, net of makeup withdrawals but including a cost for treatment of blowdown for oil-water separation, has been estimated at $0.0356/1,000 gal as described in chapter 4. Under these assumptions cooling tower installation becomes profitable and heat discharges to water are cut to zero when the effluent charge on heat reaches only $0.075/$10^6$ Btu. At the same time oil discharges decline by 77 percent, to about 0.011 lb/bbl crude, because it is assumed that cooling tower blowdown must be subject to oil-water separation. Oil recovery goes up by a corresponding amount.

These results are sensitive, however, to the estimate of cooling tower costs. Thus, for example, doubling this cost (to $0.0712/1,000 gal recirculated) raises the critical effluent charge on residual heat to $0.197/ 10^6 Btu, or by nearly three times. In this case the refiner also finds it profitable to install the high capital (waste-heat-saving) reformer when the effluent charge is $0.088, before he goes to cooling towers. (This level for cooling tower costs, however, does seem unrealistically high.) This sensitivity results from the fact that the critical balance in the decision is between the cost of waste heat discharge and the *differential* between a once-through and a recirculated system for providing cooling water. The critical level of effluent charge is defined by the equation:

$$\hat{E} = \frac{\text{critical effluent}}{\text{charge on heat}} = \frac{V(c - w) + m(w)}{H}$$

where V is cooling water volume; c is cooling tower costs; w is cost of water withdrawals; m is makeup volume for cooling tower recirculation volume, V; and H is rejected heat load.

If one takes the cooling water and makeup volumes in 1,000 gal and the heat load in 10^6 Btu, under the assumptions made in constructing the cooling tower activity, 1 unit of water carries away 0.292 units of heat and requires 0.0925 units of makeup withdrawals. Thus the critical effluent charge becomes:

$$\hat{E} = \frac{1(c - w) + 0.0925(w)}{0.292} = 3.42(c - w) + 0.316(w).$$

Assume for example that the original estimate was high and that cooling tower costs are really about 2¢/1,000 gal. Then the critical effluent charge on heat would be about 3.7¢/10^6 Btu.

Effluent Charges and Upper Limits on BOD Discharges

The response of the basic refinery with benchmark product mix to various levels of effluent charge on oxygen-demanding organics (expressed as 5-day BOD) is shown in figure 16.[7] The pattern observed includes, first,

---

[7] The figures in this chapter are generally drawn to simplify the actual results obtained from the model. Only the values of the charges actually imposed in the program and the corresponding percentage reduction or discharge level are graphed. In reality, of course, as the computer program increases the effluent charge, the changes in treatment activities, etc., are chosen as they become profitable, generally at charge levels between those for which full solutions were obtained.

The effluent charge–percentage reduction pairs are joined by straight lines as though the response between each were continuous along that line. This is not the case, but as a graphical convention it is more nearly correct the greater the number of individual changes that occur in the interval. Any resulting distortion is more than compensated for by greater clarity and ease of comparison.

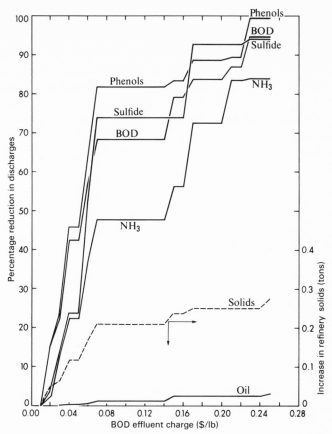

Figure 16. Response to BOD effluent charges: basic refinery, benchmark product mix.

a very rapid reduction in discharges of BOD, sulfide, and phenols over the range of $0.01/lb to $0.07/lb, at which point the refinery is "removing" almost 70 percent of its BOD load remaining after the oil-water separators and sour-water scrubbers. There is no further change until the charge reaches $0.14/lb. Then, between $0.14 and $0.15 reduction climbs to 80 percent. Between $0.15 and $0.25 there is a further steady increase to about 95 percent removal.[8] Translating the percentage reductions into actual discharges (recalling that the basic refinery discharges 9,045 lb of

---

[8] I could increase the BOD charge further and obtain further reductions, but $0.25/lb seems high enough to serve as an illustration. In addition, even though there is no actual case study against which to judge the realism of any charge level, $0.25 is higher than the figures normally discussed; these usually range from $0.05/lb to $0.15/lb of BOD.

BOD/day in the absence of controls), the following pattern is observed: by $0.03/lb, daily discharge has fallen to about 7,050 lb, by $0.07 to 2,850 lb, and by $0.15 to 1,875 lb.

In the meantime discharges of phenols have been reduced to 18 percent of their unregulated level (about 900 lb/day) at the $0.07/lb (of BOD) charge, and over the same range sulfide discharge falls by 74 percent. On the other hand, oil discharges are hardly reduced at all over the entire charge range, since they are dominated by cooling water discharge, and this is assumed to contain no other residuals and so is not affected by the BOD charge. Ammonia removal lags behind at every level of charge, but is still nearly 50 percent when the BOD charge is $0.07/lb or more.

Heat discharges do not change over the entire charge range, since no processes are affected and since cooling tower recirculation of API effluent is not chosen as the method of reacting to the BOD charge. It is quite possible that an existing refinery with *existing* cooling towers might find this an attractive route, but if the towers would have to be built simply to cope with BOD it would be cheaper to go to standard secondary and tertiary methods. On the other hand, recirculation of secondary plant effluent to the desalter unit is used. Indeed when the BOD charge is greater than or equal to $0.04/lb, 100 percent of desalter feed is being provided from this source. In addition, some streams are treated through the tertiary level and recirculated to the boiler-feed system. Up to a $0.14/lb charge, this amounts only to 10.8×10^3 gal/day (about 0.4 percent of base withdrawals for feedwater purposes). But if the charge is $0.21/lb or more, 477×10^3 gal/day (18 percent of base withdrawals) are going this route.[9]

As stressed above, however, the refinery does not simply "remove" BOD and the other residuals and have an end of it. Secondary residuals are generated in the treatment processes, and even when CO_2 and water produced in the oxidation of organics are ignored, there are some potentially significant secondary residuals generated in the course of responding to BOD charges. In particular, secondary treatment by the standard

---

[9] At this point it is worth emphasizing that I am assuming that the refinery can be designed as a grass roots plant for any level of charge. This means that some segregation of streams is possible. But I am also implicitly assuming that streams are mixed in treatment units as the basis for my cost estimates. This means that once treatment has occurred, it is unrealistic to assume that separate streams exist and can be routed in different directions. (This is distinct from the problem of economies of scale, but it is true that units designed to handle individual streams would be very costly because very small.) Thus my results must be taken to be approximate indications of levels of removal rather than hard and fast predictions.

activated sludge process results in the production of a large quantity of sludge. It has been assumed that 0.75 lb of dry sludge solids are generated per pound of BOD removed in the treatment process, and a small amount of sulfur is probably released either as H_2S or SO_2, and some hydrocarbons are vaporized. The last two of the secondary residuals are not quantitatively significant. But the stream of sludge is potentially a large problem. At a charge of $0.03/lb of BOD, for example, the dry weight of sludge solids produced would be 1,200 lb/day. But these solids would occur in the form of very dilute sludge; at 5 percent solids, the total weight of sludge generated would be about 12 tons per day. When the BOD charge is $0.22/lb, the dry weight of sludge solids is 4,800 lb/day, or over 48 tons of raw sludge at 5 percent solids. Explicit account is taken of sludge generated in the model and thickening plus incineration is required. (Below, the sensitivity of the solutions to a doubling of the cost of this requirement is investigated.) The impact of this requirement is to increase particulate generation by 1 percent at a $0.06/lb BOD charge and by 1.4 percent at a charge greater than or equal to $0.17/lb. Corresponding increases in the refinery's solid residual generation occur because of incinerator bottom ash. This increase is shown graphically in figure 16.

For comparison with the results in the basic refinery case, figure 17 shows the percentage removal of BOD, sulfide, and so on in response to the BOD effluent charge for the advanced refinery producing a lower volume of no-lead, low-octane gasoline and large quantities of desulfurized kerosene and distillate and residual fuel oils. In particular, total gasoline volume is assumed to be 60 percent of that in the benchmark product mix, while distillate fuel oil is 2.33 times as great and residual fuel oil 5.5 times as great. Both fuel oils are desulfurized in this case, where they were not in the base case. This product mix is referred to as the "high-F.O. product mix."[10]

Because the treatment and recirculation alternatives are assumed to be the same for the two cases, there are no striking differences in the patterns observed. In general, however, for the advanced refinery higher removal for all residuals but phenols is found at charges up to $0.07/lb. This results from differences in the relative volumes of various BOD-containing streams in the two refineries. An important fundamental point to note is that in both cases the costs of treatment and recirculation are low enough

[10] For detail on absolute levels of discharge for this product mix, the reader is referred to table 25.

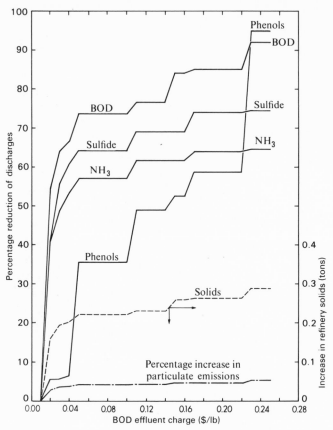

Figure 17.   Response to BOD effluent charges: advanced refinery, high–F.O. product mix.

so that changes in processes, inputs, or outputs are never chosen to help adjust to charges. It is also interesting that in this case the percentage *increase* in particulate emissions is much greater than that observed above. This is because the level of particulate emissions in the absence of BOD removal and consequent sludge incineration is very much lower for the advanced refinery, high-F.O. case. The additional pounds of particulates are roughly equal for the two cases because the original BOD loads and the overall removal are roughly the same.

It is natural at this point to ask how sensitive my results are to some of the costs I have used. In particular, it would be interesting to investigate the impact of changing the costs of the various recirculation alternatives and the cost of sludge disposal, since these are the least-documented of

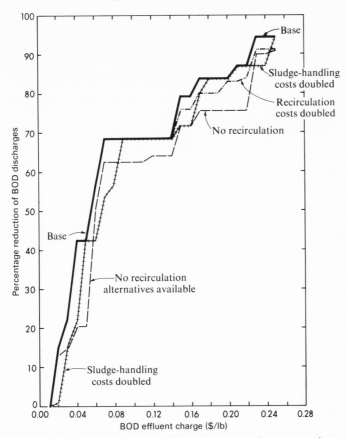

Figure 18.   Sensitivity of response to changes in assumptions concerning recirculation and sludge handling.

the costs involved. Figure 18 offers some evidence on this point: in comparison with the basic case it shows the effects of doubling (separately) the costs of recirculation and the cost of sludge disposal and of removing the recirculation alternatives entirely.[11] It is encouraging to note that the effects of doubling either the recirculation or sludge-thickening and incineration costs are quite small. Indeed the recirculation cost doubling has no discernible effect up to a charge level of $0.15/lb, and very little effect above that. Doubling the costs of sludge handling has a larger effect

[11] The recirculation activities involved are those taking effluent from a treatment process to the makeup feed side of some water use. There are three such alternatives: from oil-water separator to cooling tower; from activated sludge unit to desalter; and from activated carbon unit to boiler. The first two are assumed to cost $0.04/1,000 gal and the last $0.08/1,000 gal in the base case.

and one that is noticeable over the entire range of charges investigated. At the $0.04/lb charge level this assumption makes a difference of 20 percentage points in the level of discharge reduction adopted. For most other charges, the difference is 10 percentage points or less, and over some intervals ($0.09–$0.14/lb and $0.18–$0.22/lb) it is zero.

The elimination of the recirculation alternatives, however, makes a significant difference in the discharge reduction levels adopted; these are consistently below those of the base case for every BOD charge above $0.02/lb. The size of the difference varies from about 20 percentage points (in the area of $0.04/lb) to about 5. This is an illustration of the danger of considering only end-of-pipe treatment possibilities. Another facet of the problem is presented in figure 19, which shows that the refinery without recirculation alternatives would find it more costly to

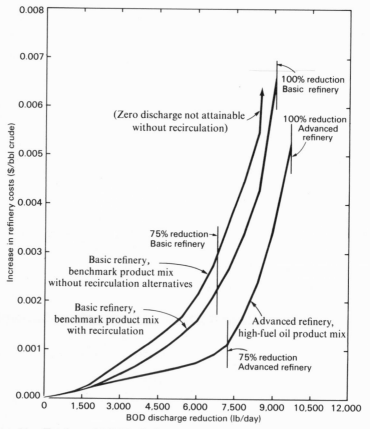

Figure 19.   Total cost of BOD discharge reduction.

meet any particular discharge constraint than would the more flexible plant.

It is worthwhile, however, to pause and consider an additional illustration of the symmetry discussed above between effluent charges and discharge constraints. Table 30 contrasts the results of placing increasingly severe constraints on BOD discharges from the basic refinery (benchmark product mix), with the results of levying increasingly stiff effluent charges

TABLE 30. Constraint–Effluent Charge Symmetry in the Case of BOD Discharges

Lb of BOD: discharge constraint →	Removal (*percent*)	→ Shadow price (*$/lb*)	Effluent charge (*$/lb*)	→ Removal (*percent*)	→ Resulting BOD discharge (*lb*)
9,045	0	0	0	0	9,045
			−0.01	0	9,045
9,000	0.5	−0.016			
8,400	7.1	−0.016			
7,800	13.7	−0.018			
			−0.02	15.1	7,680
7,200	20.3	−0.026			
			−0.03	22.2	7,032
6,600	27.1	−0.037			
6,000	33.7	−0.037			
5,400	40.4	−0.037			
			−0.04	42.5	5,199
			−0.05	42.5	5,199
4,800	47.0	−0.055			
5,200	53.5	−0.058			
			−0.06	56.5	3,933
3,600	60.2	−0.069			
3,000	66.8	−0.069			
			−0.07	68.5	2,844
			.	.	.
			.	.	.
			−0.14	68.5	2,844
2,400	73.5	−0.145			
			−0.15	79.2	1,881
			−0.16	79.2	1,881
1,800	80.2	−0.161			
			−0.17	83.8	1,464
			−0.18	83.8	1,464
			−0.19	83.8	1,464
			−0.20	83.8	1,464
1,200	86.8	−0.201			
			−0.21	87.0	1,158
			−0.22	87.0	1,158
600	93.4	−0.228			
			−0.23	94.5	492
0	100.0	−1.13			

on that residual. Shown for the constraints are the pounds per day allowed, the corresponding discharge reductions, and the shadow price (or dual value) calculated for the program for the constraint level. With the effluent charge values are shown the corresponding discharges and percent reductions. The table shows that if an effluent charge lies between two shadow price values, it will produce a discharge response also lying between the two corresponding constraints. Also, however, because of the discontinuities in marginal costs there can be close agreement between two discharge levels but a large difference between the effluent charge that produced the one and the shadow price corresponding to the other.

Finally, as an example of other possible questions that can be answered with the help of this type of model, figure 19 includes total cost functions for BOD removal for the basic (benchmark product mix) refinery with and without recirculation alternatives and for the advanced (high-F.O. product mix) refinery with recirculation. BOD discharge reduction is measured along the horizontal axis and the total (capital plus current) daily cost of this removal per barrel of crude along the vertical axis. Such a functional relation is easily obtained from the model by using discharge constraints, measuring total cost by the change in the objective function corresponding to a particular constraint level. Three features of the BOD cost functions are worth noting.

First, the advanced refinery with the high-F.O. product mix would present a regional residuals management agency with a smaller problem than the basic refinery—even though the uncontrolled level of BOD discharge is slightly higher in the former case. Thus it is obvious from the figure that the marginal and total costs of meeting a particular discharge constraint—75 percent removal, for example—are lower at the advanced refinery. And the resulting discharges are nearly the same.

Second, it is clear that the total daily costs, even of 100 percent reduction of BOD discharges, are small relative to daily costs for the refinery. Thus 75 percent reduction in BOD discharges of the basic refinery would cost only about $0.002/bbl of crude or about 0.045 percent of daily costs; complete (100 percent) reduction would be three and one-half times as expensive.[12] While these figures are subject to the same caveats that have

---

[12] Note that these costs are those of treatment and recirculation adjustments only —they do not include any effluent charge payments. Clearly when the firm is faced by an effluent charge and adjusts to equate its marginal cost of discharge reduction to that charge, unless the charge is large enough to produce zero discharges the firm will pay not only the total costs of reduction, but also an amount equal to the effluent charge times the remaining discharge. For the basic refinery, for example, 50 percent

been emphasized throughout, it seems unlikely that the cost of, say, 75 percent reduction in BOD discharges (based on post-oil-water separator load) for a grass roots refinery would be as much as 0.1 percent of daily costs.

The third interesting feature of figure 19 is the difference between the cost curves for the basic refinery with and without the availability of recirculation alternatives. The loss of flexibility imposed by assuming no recirculation is possible costs a little over $0.0003/bbl if the desired reduction is 4,500 lb, about $0.001/bbl if the desired reduction is 7,500 lbs, and becomes large without limit as zero discharges are approached. This evidence is clearly relevant to the contention that consideration only of treatment alternatives will generally lead to an overstatement of the costs of obtaining any particular desired response from dischargers and thus to an overstatement of the costs implied by any given set of ambient regional standards. The likely result of such an overstatement is a reduction in the level of ambient quality ultimately chosen by the politicians or administrators responsible for the regional decision.

OTHER WATERBORNE RESIDUALS

The refinery's responses to effluent charges (or upper limits) on BOD discharges and its response to similar actions aimed at the other waterborne residuals are basically similar so that it will not be necessary to discuss the latter in any great detail. A number of figures will be presented, however, and attention will be called to such differences as exist for phenols, sulfide, and ammonia. Oil is a special case.

Phenols

Phenols form part of the total of oxygen-demanding organics in the refinery's discharge. Hence an effluent charge on phenols produces similar responses to those observed above. But for the basic refinery there is some reliance on the cooling tower recirculation route since it is assumed that complete removal of phenols may be achieved in this way. Indeed the mix of responses varies considerably over the range of charges—from

---

reduction would be achieved in response to an effluent charge of about $0.055/lb (from figure 17). Total adjustment costs would be $0.001/bbl (figure 18), and the fee paid on the remaining discharges would be about $0.0016/bbl (4,500 × $0.055). The total cost to the firm would be about $0.0026/bbl.

secondary treatment of sprung water from the caustic scrubber at \$0.03/ lb (producing 14.5 percent discharge reduction) to extensive treatment plus the use of all three recirculation alternatives at the \$0.20/lb level with an 83.5 percent discharge reduction.

In the process of adopting these adjustments the refinery, of course, also reduces its discharges of heat, BOD, sulfide, and ammonia and produces a quantity of treatment-plant sludge that in turn leads to increased particulate emissions and solid residuals. These effects (except for the variation in heat and particulate discharges) are shown in figure 20 for the basic refinery. For comparison, the phenol reduction curve for the advanced, high-F.O. product mix is included, and the major point of interest in the figure is the great difference in results for the two refineries. For the advanced refinery, the reduction in phenols discharge is only

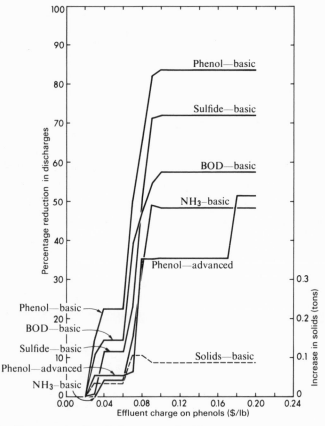

Figure 20.　Response to effluent charges on phenols.

about 5.4 percent at $0.03/lb and 51.5 percent at a charge of $0.20/lb, while the latter charge in the basic refinery is enough to produce a reduction of almost 85 percent. This difference results from the lower volumes of two important phenolic condensates in the advanced refinery. Because hydrocracking is used very heavily in preference to standard catalytic cracking, cat cracker condensates, particularly those from the cracking of coker gas oil, are considerably lower in volume in the advanced refinery.[13] In addition, only 10 percent as much sprung water, the stream with the highest phenol concentration of any in the refinery, is produced in the advanced refinery because the hydrogen-intensive processes result in generally higher-quality products that are used as reformer feed. Proportionally more of the phenol discharged from the advanced plant comes from condensate streams in which the concentrations are very low. Thus higher charges are required to spur the refinery to adopt treatment or recirculation activities to reduce these discharges.

As another indication of the kind of information obtainable from the model, figure 21 shows the total daily cost (net of any effluent charge payments) connected with levels of phenols discharge between zero and the unregulated levels for both refineries. What is most striking here, as for BOD, is the low cost of achieving even 100 percent elimination of this residual from discharged condensate, though the costs do climb very steeply as this strict control level is approached. Thus though the costs of achieving zero phenols discharges from the basic plant are over twice as high as the costs of allowing only 600 lb/day (which, itself, amounts to about 88 percent removal), these costs still only amount to less than one-tenth of 1 percent of daily costs, or less than $0.004/bbl of crude.

Sulfide

Sulfide (hydrogen sulfide dissolved in process condensates) discharges are low for both the basic and advanced refineries relative to their generation of this residual. This is because it has been assumed that sour-water stripping is applied to the condensates even in the absence of formal control actions such as charges or discharge limits. The effect of this assumption is to leave only 1 percent of the original dissolved hydrogen sulfide in the condensates and to present the refiner with the choice of flaring the

---

[13] In the basic refinery, condensates from the cat cracking of coker gas oils total 46.4×10^3 gal/day. For the advanced refinery, the corresponding figure is zero.

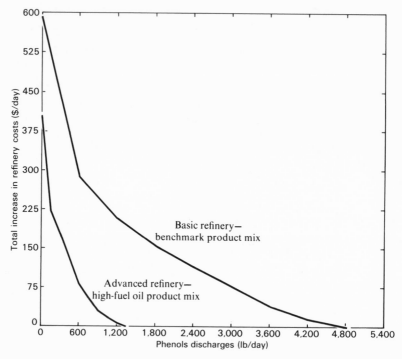

Figure 21.   Cost of achieving various levels of phenols discharges.

other 99 percent (thus producing a large quantity of SO_2) or of routing it to a sulfur recovery plant.[14] But even these low levels of discharge to the local water course may create enough nuisance problems in the vicinity of the refinery to provoke efforts at their control. Figures 22 through 24 offer some graphic evidence concerning the sensitivity of waterborne sulfide discharges to such controls. Figure 22 shows how the percentage removal of sulfide varies with the charge per pound of discharge.[15] It is clear that the advanced refinery is far more sensitive to such a charge than the basic one. This difference is accounted for by the large quantities of condensates from the hydrogen-intensive process units employed in the advanced refinery under the specifications for the high-desulfurized F.O. product mix. These condensates are very sour (that is, very high in dissolved hydrogen sulfide) and consequently large reductions in sulfide

[14] See the next chapter on airborne residuals and the comments in chapter 6 on the effect of changing the assumed price of recovered sulfur.
[15] Actually this figure was developed from solutions to the model utilizing discharge constraints; thus the corresponding "charges" are shadow prices or dual values.

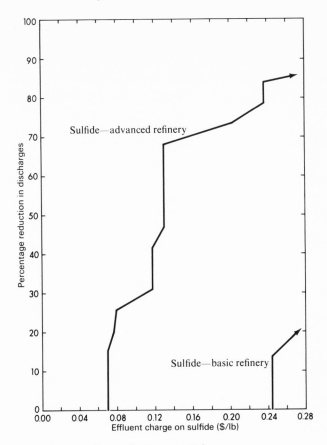

Figure 22. Response to effluent charges on sulfide.

discharges may be achieved at a relatively low cost per pound of discharge reduction.[16]

Both refineries use cooling tower recirculation of API-unit effluent as well as secondary treatment in reacting to the standards. It appears that the choice between the two methods for a particular stream is determined by the quantity of BOD in that stream. Sending a high-BOD stream to secondary treatment implies a high sludge generation rate, and hence high sludge-handling expense, while sending it to the cooling tower does not. The extent to which this choice can be made is limited, however, by

[16] Since the costs of recirculation and treatment alternatives are assumed to depend only on *volume*, the higher the concentration, the lower the cost per pound of sulfide "removed." One hundred percent of the remaining sulfide is assumed oxidized in the cooling tower and in secondary treatment.

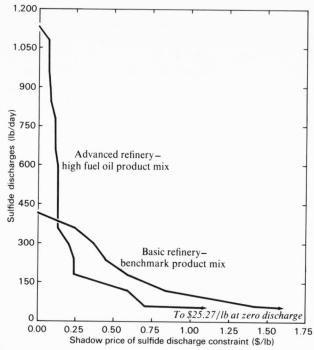

Figure 23. Sulfide discharges and related shadow prices.

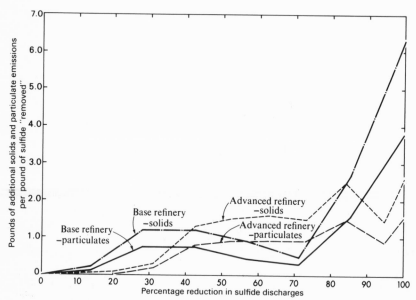

Figure 24. Implications of sulfide discharge reduction for solids and particulates.

the assumption that certain streams, notably sprung water, desalter water, and residual fuel oil hydrotreater condensates cannot be used as cooling tower makeup because of their high solids contents. (This secondary-residuals aspect of the problem is returned to below.)

Percentage removals do not really give an accurate picture of the relative impact of the two refineries on their local environment under different effluent charges. This is because of the large difference between their unregulated discharges of sulfide. For this reason figure 23 is included, in which the quantity of discharge is graphed against the effluent charge or shadow price. Though the discharges at zero charge are almost three times as great for the advanced refinery, a charge of $0.13/lb is sufficient to reduce those discharges to a level below those found in the uncontrolled situation at the basic refinery. For higher charges the advanced refinery's discharges are smaller than those of the basic plant, though they both eventually approach zero discharge at the same very high shadow price level, $25.27/lb.

Finally, it will be interesting to return to the aspect of the sulfide problem raised above in connection with the choice between cooling towers and treatment alternatives; the high cost of removal in terms of the quantities of particulates and solid residuals generated. Figure 24 is a graph of the marginal increases in solids and particulates for both refineries against the sulfide discharge reduction percentage. From the curves it is seen that over significant portions of the range of percentage reduction more than 1 lb of solid residuals and nearly 1 lb of particulate emissions are generated per pound of sulfide discharge reduction. This of course results from the use of standard secondary treatment to remove H_2S, which appears generally in very low concentrations. The lesson of this graph is that while zero discharges of waterborne residuals may be technically feasible and even cheap (in the sense that they would imply a small percentage addition to total costs), it may very well be that significant increases in other residuals problems will be implied by such an effort. Whether such increases are an acceptable "cost" can only be evaluated in the context of a regional framework in which air pollution and solid waste problems as well as water pollution are considered.

Oil

The waterborne oil residual is interesting because it represents quite a different set of discharge control problems and costs. By far the greatest part of the oil discharge from both the basic and advanced refineries results from the very low concentration of oil (5.3 ppm) assumed to be in the cooling water. This oil in turn is assumed to come from many small leaks throughout the heat exchange system. The cost of reducing the oil generation at its source by increasing heat exchange maintenance expenditures has not been estimated, and thus it is possible that the results are misleading. The major alternative open to the refiner is to go to a cooling-tower recirculation system in which it is assumed that none of the oil is lost through oxidation or evaporation, but all appears in the blowdown in much higher concentration (106 ppm, or 20 times higher) than in the once-through water. The blowdown is then subject to oil-water separation with a removal efficiency of 80 percent.[17] The shadow price (effluent charge) per pound of oil discharge corresponding to this technique is $0.62, quite a bit higher than the charge levels required to produce significant discharge reductions for the other waterborne residuals.

As one might expect, given the intimate tie between cooling load and oil discharge and the high cost of discharge reduction by treatment, the oil residual is the exception that probes another rule one is tempted to propose regarding waterborne residuals—that process changes are never chosen as part of the control strategy for them. In fact, for oil, one of the first adjustments made to lower discharge limits is to opt for the high-capital reformer to cut down on cooling required.

Two further observations are of interest here concerning the impact of the simultaneous imposition of control strategies on oil and other waterborne discharges. First, if a $0.05/lb BOD effluent charge is being levied, the response of the refinery (and the marginal cost of that response) to oil discharge limits is hardly changed at all. This is because the major source of oil discharges is not part of the BOD problem, and the sources of BOD (the process condensates) are not significant for oil discharge reduction. The only change is that some of the condensates are used as makeup to the cooling towers, and the towers themselves are introduced at a slightly lower shadow price on oil discharge. On the other hand, a $0.05/10^6 Btu effluent charge on heat has a significant effect on the response to oil discharge constraints, since the existence of such a charge makes the installation of cooling towers much cheaper in terms of the marginal costs at-

[17] I have not provided for feeding this blowdown to a secondary treatment plant and thus have as an artifact an irreducible minimum oil discharge about 20 percent of the unregulated level. Neither have I dealt with other problems in blowdown treatment, such as the presence of toxic algicide residues, heavy metal ions, etc.

tributable to oil discharge reduction. The shadow price of the cooling tower option falls from \$0.62/lb to \$0.205/lb when the heat charge is in effect.[18]

Ammonia

The refiner responds to effluent charges on ammonia in a manner qualitatively similar to that associated with sulfide (see figure 23). This is not surprising, since it has been assumed that for most process waste-water streams, the original NH_3 concentration is simply 0.75 times the H_2S concentration. The different removal levels assumed at each stage change this relation for each stream but do not change the fact that streams with high H_2S concentrations will also have high NH_3 concentrations.[19] Thus figure 25 shows that, although the uncontrolled NH_3 discharges from the advanced refinery are almost three times as great as those from the basic refinery, an effluent charge of \$0.04/lb is sufficient to induce such a dramatic reduction in the former that the relative size of the discharges is reversed. For higher charge levels, the ammonia discharges from the two refineries are nearly equal. The very large reductions at the advanced refinery result from the secondary and tertiary treatment of the process condensates from the hydrotreating of reduced crude for sale as residual fuel oil. These streams are particularly high in H_2S and NH_3 and hence are particularly cheap to treat per pound of NH_3 removed.[20] The size of the reduction in BOD discharges is about the same as the size of the ammonia reduction for each refinery at a particular charge level.

The following general observations, valid for BOD, phenols, sulfide, and ammonia, but not for oil, are repeated to summarize these sections dealing with waterborne residuals:

1. The total costs of reducing discharges of waterborne residuals, even to zero, are a very small fraction, generally less than one-tenth of 1 percent, of refinery costs.
2. Marginal costs of reducing these discharges vary widely across the four residuals, but in no case are they great enough to stimulate

[18] There is no difference between the basic and advanced refineries in any of these matters, since the same cooling tower option is available to each.

[19] This is not true for sprung water, which is assumed to contain no ammonia.

[20] This, again, depends on the assumption that treatment cost is independent of concentration and is related solely to volume.

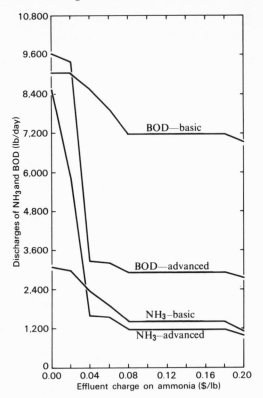

Figure 25.   Response to effluent charges on ammonia.

fundamental process changes. Treatment and recirculation alternatives are always chosen in response to the effluent charges and discharge constraints.

3. The inclusion of recirculation alternatives as options for dealing with treated condensates allows the attainment of any desired BOD discharge level at lower cost than is possible with treatment alternatives only. This is also true for the other waterborne residuals, though evidence on them has not been presented in this connection.

4. The oil residual exception suggests a fourth generalization: if a residual is present in very low concentrations in large quantities of a transport medium, treatment may be sufficiently expensive that process changes will be adopted as part of an adjustment to discharge limits or prices.

VIII

EFFECTS OF DIRECT INFLUENCES ON AIRBORNE RESIDUALS DISCHARGES

This chapter is a continuation of the investigation of the effects of such residuals management actions as effluent (or emission) charges and discharge limits. Here I turn to the refinery's airborne residuals, considering SO_2 and particulates and using the methods familiar from the last chapter to display the results. The major difference observed between airborne and waterborne residuals is that adjustments to charges or limits on the former will frequently involve fundamental process changes. Indeed, since there are no "treatment" alternatives for reducing SO_2 emissions, that residual can be dealt with only by process changes (where I include changes in input mix under this heading). The refinery has somewhat more flexibility with regard to particulates, since it has been provided with alternative processes for removing catalyst fines from the flue gases of the cat cracker catalyst regenerator and combustion particulates from boiler flues; but even here, the very highest levels of overall discharge reduction require the institution of process changes. Not surprisingly, then, the achievement of large reductions in SO_2 and particulate emissions is very expensive; instead of less than 0.1 percent of total costs as for most waterborne residuals, the costs for SO_2 are about 8 percent when the percentage reductions approach 75 percent. For particulates in the basic refinery, a 97 percent reduction in emissions would increase costs by about 1 percent, though a reduction of about 80 percent implies an increase of only 0.1 percent.[1]

[1] For the basic refinery with benchmark product mix, SO_2 emissions cannot be reduced below 54,500 lb/day, a 74.5 percent reduction from the uncontrolled level. For particulates, the lowest attainable emission rate is 870 lb/day, about 1.4 percent of the uncontrolled level.

Another consequence of the refinery's dependence on process changes in controlling SO_2 emissions is that the extent of control adopted in response to, say, a particular level of emission charge will in general depend on relative input prices, especially those for such key inputs as crude oil and fresh heat (purchased residual fuel oil). This former dependence is explored below for the case of SO_2 emissions.

EMISSION CHARGES AND UPPER LIMITS ON SO_2 DISCHARGES

Figures 26 and 27 show the response of the basic refinery to emission charges on SO_2. The first figure concentrates on the charge range of zero to \$0.26/lb of SO_2, the same range used most frequently in the last chapter for effluent charges on BOD, phenols, and so forth. It is interesting to note that over this range SO_2 emissions are much less sensitive when judged by percentage reduction in discharge than were the discharges of any waterborne residual except oil. Almost all the refinery's response is

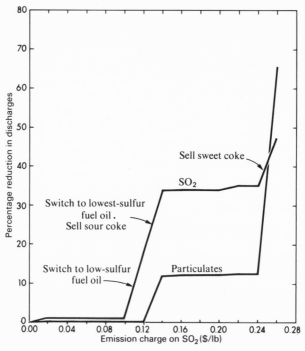

Figure 26. Response to SO_2 emission charges: basic refinery, benchmark product mix, \$0–\$0.26 range.

concentrated in the range of \$0.10/lb to \$0.14/lb, with a switch being made to low-sulfur purchased fuel (1 percent instead of 2 percent sulfur) between \$0.10 and \$0.12; and to the lowest sulfur grade (0.5 percent) between \$0.12 and \$0.14. In addition, sour coke, which is high in sulfur and is burned in the absence of SO_2 and particulate control actions, is sold when the SO_2 emission charge exceeds \$0.12/lb. This switch accounts for the reduction in particulate emissions over this charge range, since purchased residual fuel oil generates much lower particulate quantities per million Btu. There is a further jump in reductions of over 10 percentage points between charges of \$0.24/lb and \$0.26/lb, which is accomplished by selling instead of burning sweet (low-sulfur) coke. It is interesting to note the differences between the shifts away from burning

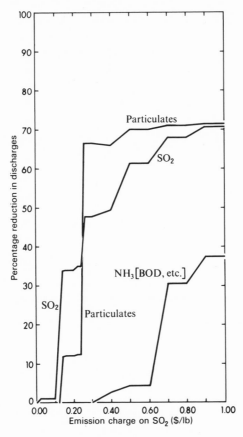

Figure 27. Response to SO_2 emission charges: basic refinery, benchmark product mix, \$0–\$1/lb range.

sour and sweet cokes. In the former case, SO_2 emissions are reduced slightly more on a percentage basis than are particulates. In the latter case, on the other hand, particulate emissions are reduced by a far greater proportion than are SO_2 emissions.

There are three small jogs in the graph of SO_2 reduction percentages: the one between zero and \$0.02/lb is caused by the adoption of the high-capital reformer, with a small resulting decrease in fresh heat input; between \$0.14 and \$0.16 several minor process shifts are introduced with the overall result of lowering catalyst regeneration very slightly at the cat cracker; above \$0.20/lb reformer polymer is burned as a substitute for purchased fuel oil.

Overall the \$0.26/lb emission charge produces a reduction of only about 48 percent in SO_2 discharges. This same level of charge on BOD stimulated a 95 percent reduction in BOD discharges.

There is no a priori reason, however, to expect that emission charges on SO_2 must be confined in the interval up to \$0.26/lb.[2] If the problem of health damage from prolonged exposure to SO_2 is shown to be serious enough, the emission charge may be set very high indeed. Hence in figure 27 the investigation is extended to a charge level of \$1/lb of SO_2. There is a slower but fairly steady reduction in discharges as the charge level goes from \$0.26/lb to \$1/lb. In the intervals \$0.40/lb to \$0.50/lb and \$0.60/lb to \$0.70/lb, however, there are significant jumps in the level of reduction (from 49.5 percent to 61.3 percent and from 61.3 percent to 68 percent, respectively). The first of these jumps is the result of three separate actions: the adoption of the high-capital coking unit; a slight switch away from high-sulfur crude; and the burning of enough desulfurized refinery gas to replace all purchased residual fuel oil. The second jump results from a much larger shift further toward charging all low-sulfur crude and a shift to a lower recycle ratio in the cat cracker, with a resulting decline in the catalyst regeneration rate. This latter set of responses also leads to a substantial decrease in the discharge of waterborne residuals, especially sulfide and ammonia, but including BOD and phenols.

Before going on to discuss the total costs of SO_2 emission reduction, three questions must be dealt with. The first is simply whether or not the response of the advanced refinery (high-desulfurized F.O. product mix) is

[2] The Nixon administration is currently in the throes of developing a proposal for a sulfur tax on fuels that would function roughly as a national emission charge on SO_2. It appears that this charge will amount to \$0.01/lb initially, rising in time to \$0.10/lb of sulfur. If 100 percent conversion to SO_2 is assumed, these levels would be equivalent respectively to \$0.005/lb and \$0.05/lb of SO_2.

substantially different from that of the basic refinery. The second concerns the effect of changes in the price obtainable for recovered sulfur,
for the results thus far have been based on an assumed price of $20/LT.
The third question is potentially of greater interest, since it involves the
relation between input prices and the response to emission control actions. This question has probably already occurred to the reader, since
it has just been observed that a shift to low-sulfur crude was part of the
response to higher levels of the SO_2 emission charge, and since a very
low cost per barrel of higher-sulfur crude was originally assumed. How
will varying the cost of the higher-sulfur crude affect the results under
discussion?

To deal first with the first two questions, figures 28 and 29 are pro-

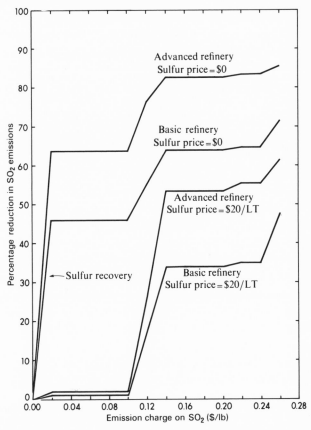

Figure 28.   Comparative sensitivity to SO_2 emission charges: basic and advanced
refineries; sulfur prices $0 and $20/LT.

vided. Figure 28 shows that, ceteris paribus, the advanced refinery responds to a given charge level with a greater percentage reduction in SO_2 emissions. It is also clear from figure 29 that the effect of variation in the price obtainable for recovered sulfur is felt entirely at the lowest charge levels; for charges above \$0.02/lb, the response of a given refinery with given product requirements is the same regardless of sulfur price. In fact, the estimates of the costs of sulfur recovery are such that the refiner will choose to carry it out rather than to flare H_2S even at a *zero* sulfur price if the emission charge on SO_2 is above \$0.0012/lb.

The major reason for the greater sensitivity to the SO_2 charge on the part of the advanced high-F.O. refinery is simply that much more residual fuel oil is purchased in this technology-product mix, and hence the impact on SO_2 emissions of going from 2 percent to 1 percent and then to 0.5 percent residual is that much greater. Since the uncontrolled emissions are roughly equal for the two cases, this impact shows up in the greater percentage reductions achieved for charge levels between \$0.10/lb and \$0.14/lb at the advanced refinery. Overall, with the \$20 sulfur price a \$0.26/lb SO_2 charge results in a reduction of 61.5 percent in the SO_2 emissions at the advanced high-F.O. refinery. This is nearly 14 percentage points greater than the response over the same range for the basic refinery.

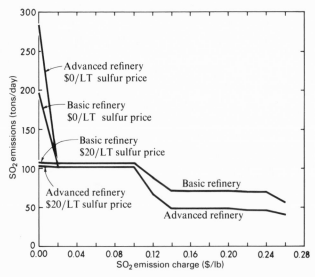

Figure 29. Response to SO_2 emission charges: basic and advanced refineries; sulfur prices \$0 and \$20/LT.

Figure 29 shows that when the price of sulfur is \$20/LT, the advanced refinery has lower SO_2 emissions for every charge level up to \$0.26/lb. When the price of sulfur is zero the advanced refinery's uncontrolled emissions are substantially larger due to flaring of the large amount of H_2S produced in hydrodesulfurization and removed from off-gases and process condensates. But since sulfur recovery is introduced at a charge level substantially less than \$0.01/lb of SO_2, the size ordering of discharges is reversed if the charge is 1¢/lb or greater.

For the basic refinery the effect of changing the assumptions about the cost of the higher-sulfur crude is shown in figure 30. Four alternative costs are considered: \$3.02/bbl (the basic assumption), \$3.22/bbl, \$3.42/bbl, and \$3.62/bbl. For the first three costs the unconstrained level of emissions is the same and the differences are in the extent to which the refiner shifts toward a 100 percent low-sulfur crude charge and in the charge levels at which the shifts are made. At \$3.62/bbl, however, it is not attractive for the refiner to charge the higher-sulfur crude even in the absence of an SO_2 emission charge. In this case reductions in SO_2 emissions are obtained by shifting to a lower-sulfur purchased heat source, selling instead of burning sweet coke and substituting capital for fresh heat inputs to major process units. When the four cases are examined for emission charges up to \$0.50/lb it is found that SO_2 emissions are, as one

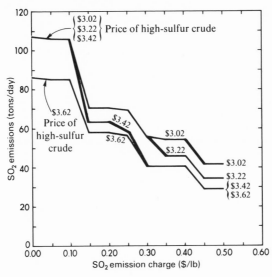

Figure 30. Variation in response to SO_2 emission charges with the price of high-sulfur crude.

would expect, always at least as low for the $3.62/bbl assumption as for any of the others. For charges over $0.30/lb, however, the emissions for the $3.42/bbl assumption are equal to those for the $3.62/bbl case. At cost assumptions of $3.02/bbl and $3.22/bbl, emissions are the same up to the $0.30/lb charge level. Above this, if the price of high sulfur crude is $3.22/bbl, SO_2 emissions are significantly lower.

In order to make clear the part played in this pattern by shifts away from the higher-sulfur crude, heavy lines denote the segments of the curves in figure 30 that reflect such shifts. From these it is clear that the effect of increasing the assumed cost of the higher-sulfur crude is to lower the charge level at which shifts away from that crude begin.

A final question concerning SO_2 discharge reductions is that of cost, and in figure 31 total reduction costs (per day and per barrel of crude) are

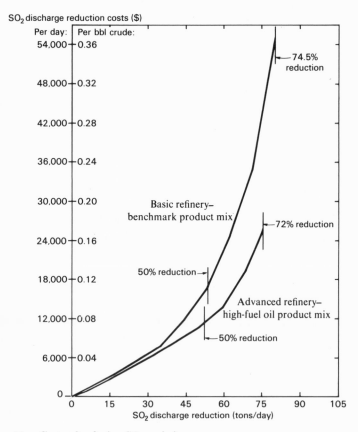

Figure 31.   Costs of reducing SO_2 emissions.

graphed as a function of the tons of discharge reduction for the basic and advanced refineries with the sulfur price assumed to be \$20/LT. From this figure it is clear that large reductions in SO_2 emissions are very expensive propositions. For the basic refinery, for example, the greatest reduction possible within the set of constraints on production, and so forth, is 74.5 percent, from 107.2 tons/day to about 27.3 tons/day. The cost of this reduction would be about \$54,000/day (or over \$0.36/bbl). For the advanced refinery the corresponding figures are a 72 percent reduction, from 104.5 tons/day to 29.2 tons/day, at a total cost of about \$25,500/day or \$0.17/bbl.

Although the maximum percentage reductions and minimum SO_2 discharges achievable by the two refineries are roughly the same, the sources of the "uncontrollable" minimum emissions are quite different.[3] In the basic model by far the largest part of these emissions comes from the cat cracker catalyst regeneration unit (68.9 percent). The gas desulfurization and sulfur recovery units account for another 22.6 percent, with 88 percent of this coming from the sulfur recovery plant. Combustion in plant boilers produces only 8.5 percent of the minimum emissions. At the advanced refinery, on the other hand, there is no catalyst regeneration carried out, and minimum SO_2 emissions are accounted for largely by gas desulfurization and sulfur recovery (69.2 percent). Sulfur recovery alone produces over 63 percent of the total. Combustion produces the other 30.8 percent, and this in turn comes very largely from the burning of 0.5 percent sulfur residual fuel oil. (There is not enough refinery gas produced at the advanced refinery, under the high-F.O. product mix constraints, to provide for all the fresh heat requirements.)

EMISSION CHARGES AND UPPER LIMITS ON PARTICULATE EMISSIONS

As noted, achieving the highest feasible reductions in particulate emissions from the refinery involves the institution of fundamental process changes. For example, for the basic refinery the step from 97.2 percent to 98.1 percent reduction requires going to the high-capital reformer and has a shadow price of over \$30/lb of particulates. But substantial reductions below these upper limits are possible at costs more like those found

---

[3] I put "uncontrollable" in quotes to draw attention to the fact that there may be control methods available that are not in the model; an example is the sulfur recovery plant tail gas, for which a second sulfur recovery plant, based on a different process, is a possibility I have not included.

for waterborne residuals. Thus figure 32 shows that a reduction of 78 percent from the uncontrolled level is stimulated in the basic refinery by a charge of $0.01/lb, and a charge of $0.02/lb produces a reduction of 82.5 percent. The very high levels of reduction at low emissions charges are produced by the application of standard stack gas-cleaning equipment to the cat cracker catalyst regenerator and to the boiler burning petroleum coke. Specifically, the refinery's responses and the necessary charge levels can be summarized as follows:

Charge level ($/lb)	Unit affected	Equipment chosen
0.0052	Coke-burning boiler	Low-efficiency cyclone
0.0063	Coke-burning boiler	Medium-efficiency cyclone
0.0072	Catalyst regenerator	Low-efficiency cyclone
0.008	Coke-burning boiler	High-efficiency cyclone
0.0149	Catalyst regenerator	Medium-efficiency cyclone
0.1067	Catalyst regenerator	High-efficiency electrostatic precipitators

The pattern of response for the advanced refinery is similar, though the removal levels attained at specific charges are uniformly lower. In the advanced refinery, however, the reductions depend much more heavily on the control of boiler emissions, since cat cracking is far less important than in the basic refinery (and indeed goes to zero as the lowest attainable emission level is approached). Coke is the largest single source of particulates in the uncontrolled case, accounting for over 81 percent. Almost all the rest results from combustion of purchased residual fuel oil. At the 1,500-lb discharge limit, the purchased residual fuel oil becomes the most important source (about 67 percent), the rest coming from coke combustion. The reductions in total discharges have been accomplished by a reduction in the quantity of coke burned (to 20 percent of the quantity in the unconstrained case) and by the application of gas-cleaning devices of the highest available efficiency to the boiler flues.

As shown in figure 33 for the basic refinery, the adoption of these adjustments implies the generation of a significant solid residual load, essentially equal in weight to the reduction in fly ash emissions and, it is assumed, costing $2 per ton for disposal to landfill.[4] At very high charges,

---

[4] I do not distinguish between particles of different sizes in the emissions, though this is important for such considerations as the location of fallout and (probably) health effects. In general it is easier to remove the large particles from the flue gases, and the greater the percentage removal achieved, the greater the proportion of very small particles in the remaining discharges.

of course, the additional reduction in discharges is achieved through fuel substitution and process changes. No additional solids load is generated by these moves; indeed, total solids are decreased sharply at very high reduction levels by shifts to gaseous fuels.

In figure 34 the cost of attaining different particulate emission levels is graphed against the level achieved. For the basic refinery this function is very flat down to about 12,000 lb/day (81 percent discharge reduction), at which level it turns very steeply upward. The difference in cost between 12,000 lb and 6,000 lb of particulate emissions is $1,880/day. At 12,000 lb of emissions the shadow cost is $0.0052/lb, while at 6,000 lb it is $0.209/lb. But even after this sharp change the total costs are of the same order of magnitude as those observed in chapter 6. For levels of discharge below 3,000 lb/day, however, the cost curve becomes nearly

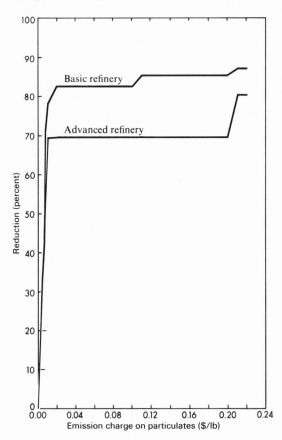

Figure 32. Response to an emission charge on particulates.

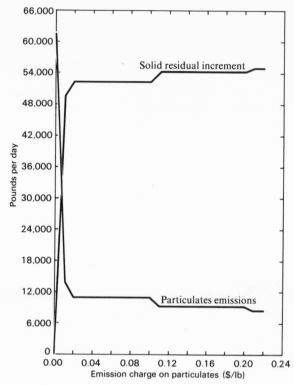

Figure 33. Particulate emissions and solid residuals generation versus an emission charge on particulates.

vertical. To achieve 98.2 percent reduction (emissions of about 1,200 lb/day) costs over $18,600/day, which is about 3 percent of total costs in the benchmark case (with no controls on any residuals discharges), or $0.12/bbl of crude charged.

By way of comparison figure 34 also shows the costs of achieving given discharges for the advanced refinery with high-desulfurized F.O. product mix. The pattern is very similar to that found for the basic refinery, but it is considerably cheaper to achieve any particular level of emissions. The total costs of given *percentage reductions* in discharges are roughly equal for the two refineries, at least for levels above 50 percent.

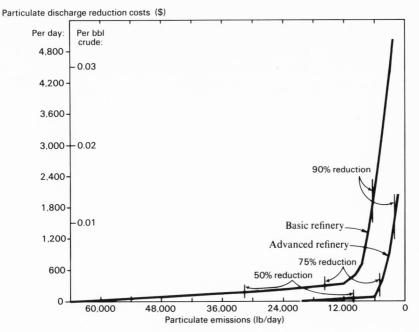

Figure 34. Costs of attaining different levels of particulate emissions.

IX

CONCLUSION: A SUMMARY AND EXAMPLES OF APPLICATION AND EXTENSION

This study has involved two levels of argument that imply two ways of looking at the results. On one quite general level the argument has been that if one's interest is in industrial residuals or industry's response to public residuals management actions, the approach that rests on "generation coefficients" tied to sales, employment, input, output, or whatever, and on "treatment" cost functions, is simply inadequate to the task of providing information and predictions under a variety of possible future circumstances and for a variety of policy instruments. The generation coefficient depends on too many other things being equal; and treatment cost functions ignore the possibilities for process change that may be the cheapest response. To support this argument I have used the petroleum refinery model to show that residuals generation varies substantially with changes in such factors (called "indirect influences") as process technology, product mix, product quality specifications, and, of course, input quality. These results were contained in chapter 6. I also showed that nontreatment alternatives for reducing residuals discharges can make a big difference. For some residuals, especially SO_2, they are essentially the only methods available (chapter 8). For others—for example, BOD, phenols, sulfide, and ammonia—the costs involved in attaining any particular level of discharge are lower when nontreatment alternatives are considered (chapter 7).

On another level, however, I was inevitably interested in the petroleum refinery model for its own sake, for what it could reveal about the likely response of a major industry to a variety of residuals management measures under different assumed sets of indirect influences. I have inserted a number of caveats to the effect that the numerical results should be regarded with caution, and that my major interest was in the model itself and the more fundamental line of argument. Nonetheless a summary of the results in the first part of this chapter provides an interesting comparison of the relative impacts of the many influences explored.

In the second part of the chapter I return to the broader theme of the utility of the model by discussing three applications. Two of these are basically similar to experiments reported above, but they have been tied to questions that might be asked by those with policy responsibilities. The third application is related to the regional model framework discussed in the introduction and involves the possibility of condensing large models, such as the refinery model, in order that a number of such models can be combined into a single large program without going beyond the effective size limits of available solution algorithms.[1]

A SUMMARY OF RESULTS

Table 31 is a summary of most of the results reported above for variations in residuals discharges with changes in a number of direct and indirect influences. The table is arranged to show percentage changes in residuals discharges and total fresh heat applied and by-product sulfur recovered, resulting from specified changes in constraints or prices. This makes it easy to see that a change in gasoline quality requirements, for example, produces changes in residuals discharges of the same order of magnitude as the imposition of a significant effluent or emission charge. The table as a whole, then, may be viewed as strong evidence for the argument that indirect influences should not be ignored.

In assessing what has been revealed about the refinery in particular, I suggest the following observations based on the table:

1. Certain indirect influences have very large effects on all or most of the residuals discharges. This is true of water costs, product quality specifications, and product mix requirements. The impacts of crude sulfur content are only slightly smaller and less widespread.

[1] In a regional context it is reasonable to expect that fairly detailed models of the major residuals dischargers would be desired, but that it might very well be necessary to settle in practice for generation coefficients and perhaps a limited number of options for controlling discharges for the vast majority of industrial sources.

TABLE 31. The Impact of Direct and Indirect Influences on Residuals Discharges

	Percentage changes in residuals discharges								Percentage change in total heat applied
	Heat to water	Sulfide	Phenol	Oil	BOD	NH$_3$	SO$_2$	Particulates	
For basic refinery									
Tripling cost of water withdrawals	-100	-99.2	-84.0	-77.0	-46.1	-31.2	neg.	0	0
Doubling cost of fresh heat ($0.20 to $0.40/10^6 Btu)	-1.8	0	0	-1.8	0	0	-3.2	+116.0	-4.0
Tripling cost of fresh heat ($0.20 to $0.60/10^6 Btu)	-2.8	0	0	-2.6	0	0	-4.2	+116.0	-5.6
Increasing average sulfur content of crude by 50% (0.4 to 0.608%)	+1.2	+54.4	+2.6	+1.2	+2.4	+54.8	+17.2	+1.4	+1.2
For advanced refinery									
To modified benchmark product mix with no-lead/low-octane gasoline[b]	+8.0	+33.0	+3.0	+6.1	+1.6	+55.0	+5.4	+0.2	+11.8
To modified benchmark product mix with no-lead, high-octane gasoline[b]	+11.3	+66.7	-8.8	+10.2	-7.9	+90.0	+12.2	-2.6	+22.6
To low gasoline (no-lead, low-octane); high fuel oil; all fuel oils + kerosene desulfurized[d]	-16.4	+167.0	-73.5	-18.4	+1.6	+185.0	-2.7	-68.1	+25.6
Effluent emission charges[e]									
Phenols, basic	-0.4	-11.5	-22.4	-0.3	-14.7	-4.4	0	+0.2	0
Phenols, advanced[d]	0	0	-5.4	0	-1.0	0	0	neg.	0
BOD, basic	0	-23.8	-45.9	-0.4	-42.5	-22.7	0	-0.7	0
BOD, advanced[d]	0	-64.2	-35.5	-0.8	-73.5	-56.9	0	+4.0	0
BOD, no recirculation, basic	0	-13.7	-22.8	-0.1	-20.3	-6.6	0	+0.4	0
NH$_3$, basic	0	-31.2	-8.8	-0.3	-9.0	-31.1	0	+0.2	0
NH$_3$, advanced[d]	0	-88.7	-4.7	-0.3	-66.3	-81.5	0	+3.5	0
SO$_2$, basic	-1.0	0	0	-1.0	0	0	-1.0	-0.1	-1.6
SO$_2$, advanced[d]	-2.2	0	0	-2.1	0	0	-1.8	-0.5	-2.3
Particulates, basic	0	0	0	0	0	0	0	-82.5	0
Particulates, advanced[d]	0	0	0	0	0	0	0	-69.5	0

	Percentage change in by-product sulfur	Cost or gain ($/bbl crude)			Percentage change in cost
		Cost of discharge reduction	Discharge fee paid	Total cost increase	
For basic refinery					
Tripling cost of water withdrawals	0	—	—	0.031	0.70
Doubling cost of fresh heat ($0.20 to $0.40/10^6 Btu)	0	—	—	0.054	1.23
Tripling cost of fresh heat ($0.20 to $0.60/10^6 Btu)	0	—	—	0.098	2.24
Increasing average sulfur content of crude by 50% (0.4 to 0.608%)	+46.3	—	—	a	a
For advanced refinery					
To modified benchmark product mix with no-lead/low-octane gasoline[b]	−2.8	—	—	0.076[c]	1.72[c]
To modified benchmark product mix with no-lead, high-octane gasoline[b]	+39.4	—	—	0.206	4.65
To low gasoline (no-lead, low-octane); high fuel oil; all fuel oils + kerosene desulfurized[d]	+88.0	—	—	0.071	1.60
Effluent emission charges[e]					
Phenols	0	0.0002	0.0013	0.0015	0.03
Phenols, advanced[d]	0	neg.	0.0004	0.0004	0.01
BOD, basic	0	0.0007	0.0017	0.0024	0.05
BOD, advanced[d]	0	0.0010	0.0009	0.0019	0.04
BOD, no recirculation, basic	0	0.0003	0.0024	0.0027	0.06
NH3, basic	0	0.0002	0.0007	0.0009	0.02
NH3, advanced[d]	0	0.0010	0.0005	0.0015	0.03
SO2, basic	0	0.0003	0.0707	0.0710	1.68
SO2, advanced[d]	0	0.0005	0.0684	0.0689	1.57
Particulates, basic	0	0.0023	0.0037	0.0060	0.14
Particulates, advanced[d]	0	0.0006	0.0020	0.0026	0.06

[a] Cost here depends on the assumed price of high-sulfur crude. At $3.02/bbl, it is $0.109/bbl.

[b] Product mix requirements: premium gasoline 35,100 bbl/day; regular gasoline 51,150 bbl/day; kerosene 16,500 bbl/day; distillate fuel oil 15,000 bbl/day; residual fuel oil 3,000 bbl/day.

[c] Gain instead of cost.

[d] Advanced refinery with product mix requirements: premium gaso-

line (92 octane) 21,060 bbl/day; regular gasoline (90 octane) 30,690 bbl/day; kerosene (desulfurized) 16,500 bbl/day; distillate fuel oil (low sulfur) 40,500 bbl/day; residual fuel oil (desulfurized) 16,500 bbl/day.

[e] In each case, $0.05/lb. Such charges produced no effects in the cases of oil and sulfide, nor did a $0.05/10^6 Btu charge have any effect on heat discharges.

2. Changing the cost of water withdrawals (and pretreatment) affects the waterborne residuals discharges dramatically but has no impact on discharges to the atmosphere. On the other hand, the effects of changes in the price of fresh heat are almost all confined to the SO_2 and particulate residuals.

3. While direct comparisons of the costs implied by the various indirect influences are made difficult by the very nature of the changes specified, a few are possible. Clearly a tripling of the cost of fresh heat would be a more serious matter to the refiner than a tripling in water withdrawal costs. Large changes in product mix, including greatly increased production of desulfurized distillate and residual fuel oil, also imply large costs.[2] Indeed, it is rather striking to find that all the costs per barrel are of roughly the same order of magnitude.

4. One of the largest effects of requiring a product mix heavy in hydrodesulfurized streams is the increased recovery of by-product sulfur. Sulfide and ammonia discharges also increase significantly.

5. Effluent or emission charges on different residuals produce widely varying effects on discharges and costs. To facilitate comparison, the table is based on a constant $0.05/lb charge for each of five residuals.

   (a) This charge level, when applied to oil and sulfide discharges, has no effect at all; the refiner simply chooses to pay the resulting small fee. The same is true for a $0.05/$10^6$ Btu charge for waste heat discharge.

   (b) For phenols, BOD, and ammonia, on the other hand, the $0.05/lb charge is sufficiently high to evoke significant discharge reductions for most of the waterborne residuals. (Increases in particulates discharges result from the incineration of sludge.)

   (c) There is no consistent pattern to the relative sensitivity of the two different refineries. The basic model is more sensitive to the charge on phenols, while the advanced model reacts more sharply to the charges on BOD and ammonia.

   (d) The assumption that none of the recirculation alternatives are

[2] This finding implies that prices will in fact have to shift if the rate of return to refining capital is not to fall significantly in the event that product mix changes become necessary because of government policy. If demands shift—for example, because of increasing jet travel—the product mix changes will be elicited in the classic market manner. See below for a discussion of even larger changes in the production of jet fuel.

available makes a large difference in discharge reductions attained under the $0.05/lb charge. Much less difference is produced, of course, in the costs, since the effluent charge must be paid on the pounds not "removed."

(e) Both refineries respond to a $0.05/lb charge on particulate emissions with significant discharge reductions; achieving these reductions has no side effects on any other refinery residual (except solids, which are not included here). Total payments for the basic refinery in this situation are somewhat higher than for the waterborne residuals.

(f) SO_2 is, as observed in chapter 8, the refiner's most expensive problem. Faced with a $0.05/lb charge on its emission, he opts for very small discharge reductions and pays very large discharge fees, under both product-technology assumption sets.

FURTHER EXAMPLES OF APPLICATION TO POLICY QUESTIONS

As a way of emphasizing my view of the potential utility of the industrial response model, I apply it in this section to two hypothetical policy questions. First, I investigate what would happen to refinery residuals discharges (in the absence of direct controls or charges) should external combustion (steam) automobiles burning desulfurized kerosene become a significant factor in the national transportation sector. Specifically I assume that desulfurized kerosene production would be raised to 48,000 bbl/day, using advanced refining technology.

As a second example, the model has been asked to predict the costs (net of any effluent charge payments) incurred by the refinery in meeting various uniform percentage discharge reductions for all the primary residuals (heat, BOD, sulfide, phenol, oil, SO_2, and particulates).[3]

Greater Kerosene Production

Over the last few years there have been intermittent flurries of public excitement when we have been promised that the steam car is about to stage a comeback and that when it does our urban air-pollution problems will be over. Although each of these predictions has proved premature,

---

[3] This policy is simply used as an illustration. I do *not* mean to imply approval of it.

the prospects for the steam car have been thoroughly investigated,[4] and it appears that an external combustion engine probably is technically feasible and probably could be designed to be a much smaller pollution problem than the internal combustion engine. So it does not seem far-fetched to inquire of the model what would be the effect on *refineries*, and especially on refinery discharges, of the widespread adoption of such an engine assumed to burn desulfurized kerosene. I have not actually estimated future demands for fuel under these conditions; rather I have confined myself to raising the required level of desulfurized kerosene output to 48,000 bbl/day for the advanced refinery (32 percent, by volume, of crude charged), at the same time lowering total gasoline production requirements to 51,750 bbl/day (21,060 bbl of premium and 30,690 bbl of regular).[5] The results of this experiment appear in table 32, where they are compared with the results for three other possible product configurations: (1) the modified benchmark mix with high-octane, no-lead gasoline; (2) the modified benchmark mix with low-octane, no-lead gasoline; and (3) the familiar high-desulfurized F.O. product mix.

The first line in this table is the objective function value per barrel of crude; that is, the costs of crude, processing, required treatment, and so forth, net of receipts for straight-run gasoline, by-product sulfur, refinery gases, and polymer. The cost implications of going to no-lead gasoline (columns 1 and 2) have already been discussed, but it is worth pausing a moment to put the cost figures in the last two columns in focus since the product mixes are so different from the benchmark case. The net cost for the basic refinery, benchmark case, was $3.48, and the return on sales of major products, 21.1 percent.[6] From the figures below, it is apparent that, although net costs have changed very little from the benchmark case, the return on sales of major products is about one-third lower when these sales are valued at the same prices used in chapter 5. In order to maintain

[4] See Robert U. Ayres and Richard P. McKenna, *Alternatives to the Internal Combustion Engine: Impacts on Environmental Quality* (Johns Hopkins University Press for RFF, 1972).

[5] Another contender for the role of successor to the piston internal combustion engine is the Wankel rotary. Apparently this engine runs on very low-octane gasoline, and this refinery product mix can be seen as providing for a combination of the Wankel *and* external combustion engines, since the gasoline produced is no-lead, low-octane.

[6] In table 19 I reported the refiner's overall rate of return—i.e., net income (total receipts minus total costs) divided by total receipts—which was 17.4 percent. The return on sale of major products is calculated as receipts from the sale of major products minus net costs over net costs, when net costs equal total costs plus credit for by-product sales.

TABLE 32. Impact of Doubling Jet Fuel Production

	Modified benchmark product mix, high-oct., no-lead gasoline (1)	Modified benchmark product mix, low-oct., no-lead gasoline (2)	High-F.O. product mix, low-oct., no-lead gasoline (3)	High-jet fuel product mix, low-oct., no-lead gasoline (4)
Net costs (*$/bbl crude*)	3.703	3.364	3.511	3.524
Barrels crude charged/day	191,700	150,000	150,000	150,000
Total gasoline (*bbl*)	86,250	86,250	51,750	51,750
Total kerosene (*bbl*)	28,650	16,500	16,500	48,000
Jet fuel (*bbl*)	—	720	16,500	48,000
Total distillate fuel oil (*bbl*)	15,000	15,000	40,500	15,000
Low-sulfur distillate fuel oil (*bbl*)	6,930	7,150	40,500	15,000
Total residual fuel oil (*bbl*)	3,000	3,000	16,500	6,000
Desulfurized residual (*bbl*)	—	—	16,500	—
H_2 produced (*lbs/bbl*)	0.107	0	0.799	1.809
Heat purchased (*10^6 Btu/bbl*)	0.340	0.292	0.447	0.428
Total fresh heat (*10^6 Btu/bbl*)	0.487	0.445	0.501	0.588
Sulfur recovery (*LT/day*)	59.19	41.25	79.83	63.24
Straight-run gasoline sold as petrochemical feed (*bbl/day*)	44,100	17,100	21,000	21,000
Coker (*bbl feed/bbl*)	0.138	0.133	0.051	0
Reformer (*bbl feed/bbl*)	0.243	0.193	0.253	0.275
Cat cracker (*bbl feed/bbl*)	0.295	0.369	0.004	0.012
Alkylation (*bbl produced/bbl*)	0.079	0.083	0.048	0.028
Hydrocracker (*bbl feed/bbl*)	0.142	0.056	0.129	0.282
H-oil (*bbl feed/bbl*)	0	0	0	0.115
Hydrotreating (*bbl feed/bbl*)	0.179	0.139	0.370	0.271
Residuals				
Heat to water (*10^6 Btu/day*)	66,200	50,400	39,000	43,200
(*10^6 Btu/bbl*)	0.346	0.336	0.260	0.288
Sulfide (*lb/day*)	960	600	1,200	1,050
(*lb/bbl*)	0.005	0.004	0.008	0.007
Phenols (*lb/day*)	5,940	5,250	1,350	1,200
(*lb/bbl*)	0.031	0.035	0.009	0.008
Oil (*lb/day*)	10,300	7,800	6,000	6,600
(*lb/bbl*)	0.054	0.052	0.040	0.044
BOD (*lb/day*)	11,100	9,600	9,600	3,150
(*lb/bbl*)	0.058	0.064	0.064	0.021
NH_3 (*lb/day*)	7,280	4,650	8,550	7,500
(*lb/bbl*)	0.038	0.031	0.057	0.050
SO_2 (*lb/day*)	308,000	226,500	209,100	192,450
(*lb/bbl*)	1.607	1.510	1.394	1.283
Particulates (*lb/day*)	78,400	63,150	20,100	22,950
(*lb/bbl*)	0.409	0.421	0.134	0.153

the same rate of return, prices for jet fuel and probably for distillate and residual fuel oil will have to increase substantially.[7]

	Receipts ($/bbl crude)	
	High-F.O. mix	High-jet-fuel mix
Premium gas ($6.22/bbl)	0.875	0.875
Regular gas ($5.38/bbl)	1.100	1.100
Kerosene ($4.96/bbl)	0.545	1.560
Distillate fuel oil ($4.60/bbl)	1.243	0.460
Residual fuel oil ($3.10/bbl)	0.341	0.124
Total	4.104	4.119
Less net costs	−3.511	−3.524
Net income	0.593	0.595
Rate of return on sales of major products	14.4%	14.4%

In looking over the product mix figures in table 32, the reader may be bothered by what appear to be very low yields of liquid products per barrel of crude, particularly for columns 1 and 2. The volume yield is 92.3 percent for column 1 and 92 percent for column 2. The high-jet-fuel column (4) has an apparent yield of 94.5 percent, while for the high-F.O. column, the yield of products in the table is about 97.4 percent. This apparent discrepancy is simply the result of the particular processes chosen in these cases and the mix of by-products thus produced. In each of the solutions summarized in columns 1 and 2, there is significant catalytic cracking carried out. This implies a large production and combustion of catalyst coke and also significant production of refinery gas. In column 4, the H-oil unit is used, with a consequent production of pitch. When these three materials are converted to volume equivalents and included in product streams, the yields for the four cases are all approximately 98.5 percent on a *volume* basis.

The residuals-discharge implications of the high-jet-fuel product mix requirement follow, of course, from the types of processing used to meet that requirement set. In general, these discharges are similar to those observed for the high-F.O. refinery (column 3). The only significant difference is in BOD; the fact that residual fuel oil is not desulfurized in the jet fuel mix accounts for this difference, the condensates from this process

[7] The prices I use here exaggerate the extent to which prices will have to rise, since I have not differentiated between kerosene and jet fuel.

having been assumed to be very high in BOD. By the same token, the lower overall level of hydrotreating use causes sulfide and ammonia discharges to be lower in column 4 than in column 3. The differences between the two refineries in the use of the cat cracker and coker and in total fresh heat requirements balance out in such a way that SO_2 is lower for the jet fuel mix, though particulates are higher. Both differences are slight, however.

Both the high-F.O. and high-jet-fuel product mixes imply lower discharges than the higher-gasoline-volume mixes, nearly across the board. The only exceptions are sulfide and ammonia and the cause of this is, of course, the very large amount of hydrotreating done to meet the desulfurized product requirements in the former cases. Thus it does not appear that a shift to dramatically higher turbine fuel production is going to result in trading smaller problems on city streets for larger problems in the vicinity of refineries. Indeed the model predicts that both problems will be smaller if the internal combustion engine is exchanged for the external combustion engine.

Costs of Uniform Percentage Reduction in Residuals Discharges

As a final numerical example of the kind of information that can be obtained from the petroleum refinery model in isolation, consider the question of the costs of achieving various uniform percentage reductions in *all* residuals discharges. The total costs of any feasible combination of discharge reductions may easily be found using this model; the only limits are the researcher's imagination and the computer budget. How such repeated solution might be used in constructing "condensed" models for use in large-scale regional models is discussed below.

Figure 35 shows how total daily costs vary with the uniform percentage reduction in all residuals discharges for two combinations of technology and product mix. Up to a reduction of 25 percent, total costs are very small—less than \$0.05/bbl. For 50 percent reduction, costs rise to about \$0.13/bbl for the basic refinery but to only about \$0.09/bbl for the advanced. At 70 percent reduction, costs for the basic refinery are nearly \$0.30/bbl, almost twice those for the advanced plant. As already observed, it is infeasible to reduce SO_2 and oil discharges below certain fairly high levels. (The particular levels associated with the individual residuals and particular technology-product mix assumptions are discussed in chapters 7 and 8.) For this reason, I have had to adopt one or

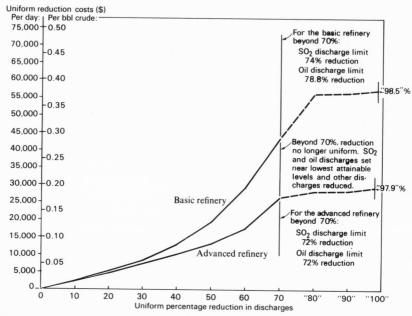

Uniform reduction costs ($)

Figure 35. Cost of uniform reduction in all residuals (except solids).

another unsatisfactory conventions in tracing out a cost curve for levels of reduction much above 70 percent. As explained on the figure, my method has been to choose reductions close to the limits of feasibility for SO_2 and oil and then to continue reducing all other discharges, leaving the SO_2 and oil limits fixed. This, of course, drastically alters the nature of the curves above the 70 percent level, and I indicate this by the use of quotation marks and dotted lines. Thus when the "98.5" percent limit on uniform discharge reductions is reached by the basic refinery, SO_2 discharges are actually at 26 percent of the uncontrolled level, oil discharges at 21.2 percent of their uncontrolled level, and all other residuals discharges at no more than 1.5 percent of their base quantities.[8] The basic lesson of this upper end of the graph is that total costs are very heavily influenced by the cost of SO_2 discharge reduction. The increment to this cost attributable to the reduction of all other residuals discharges to very low levels is hardly noticeable. This is certainly additional evidence, were any necessary, for the importance of research into treatment methods for the removal of SO_2 from stack gases. A breakthrough in this area would

---

[8] Solids are not included here. They necessarily increase substantially over this range of uniform removals.

have an enormous impact on the overall costs for achieving significantly higher levels of ambient atmospheric quality.

The last two and one-half chapters have concentrated on what the model can do by way of reflecting information and constraints on technology, product mix, input cost, and input quality while exploring the impacts of effluent charges and discharge constraints. A discussion of the model's applications in regional models and a few recommendations for further research in this area follow.

INDUSTRIAL MODELS IN THE REGIONAL FRAMEWORK

In a general way the introductory chapter covered how information about industrial response to residuals management action would fit into a larger regional model that would also reflect the diffusion, dispersion, and transformation processes of the natural world and some basis for evaluating the resulting ambient concentrations, species populations, and so on.[9] Clearly, however, it would not be possible to combine very many models the size of the petroleum refinery model in the generation and discharge section of such a regional model before reaching the limits of reliability of currently available linear programming solution algorithms.[10] One possible way around this problem is the construction of condensed versions of the full plant models. The condensation process would consist of:

1. A choice of a limited number of important inputs, outputs, and residuals that would determine the number of rows in the new model.
2. Repeated solution of the larger model, probably using different discharge constraint sets and including any desired variation in input or output constraints.
3. The characterization of each solution as a vector with entries in the rows that have been determined. These entries would be reduced to some standard unit base—most conveniently for the refinery, the barrel of crude charged.

[9] Two broad kinds of evaluation are relevant. In one kind, damage functions translate these physical effects into dollars. In the other, the physical effects are constrained and the "evaluation" consists of the assessment of penalties for nearing or exceeding these constraints.

[10] The petroleum refinery model has over 200 rows and the algorithm I have used seems to become unreliable because of rounding errors in matrix inversion when row size is in excess of 1,500.

4. The expression of the objective function value from the larger model's solution in terms of the same standard unit chosen.
5. The addition to the set of summary vectors just derived of the necessary explicit discharge activities to which trial effluent charges may be attached (as described above).

The definition of the objective function entry for these vectors and the types of stimuli to be applied to the condensed vectors are intimately related. As a general rule, if an activity is priced in the full-scale model, and the price-quantity product is left in the objective function value attributed to the condensed model vector, the condensed model will not provide a valid comparison if a new price is applied to that activity. In the example below, since the price-quantity products for purchased heat are in the objective function values, a legitimate comparison of activities in the condensed model could not be obtained by providing a heat supply activity with some price attached and having the program choose among vectors partly or wholly on the basis of their purchased heat requirements. Care obviously must be taken in the construction stage to anticipate the stimuli to be used in operation.

This process can be illustrated by considering four solutions to the basic refinery model: one in the absence of all residuals restrictions, one including a limit of 7,200 lb on BOD discharge, one with a limit of 4,200 lb on BOD discharge, and one with a limit of 1,800 lb. For simplicity, the summary vectors are constructed with only 14 rows covering crudes charged, production of regular and premium gasoline with lead additions allowed and at standard octanes, and residuals discharged. These vectors are constructed by dividing the appropriate solution value by 150,000 and are summarized in table 33.

Compare vectors (1) and (2) in the presence of an effluent charge (e_B) on BOD discharges. Since no other effluent charges are being levied, a program called on to minimize the value of the objective function, subject to some minimum required gasoline production, would decide between vectors (1) and (2) on the following basis. When

$$e_B(0.0603) + 3.4813 > e_B(0.0480) + 3.4816$$

choose vector (2). But this condition reduces to

$$e_B(0.0603 - 0.0480) > 3.4816 - 3.4813$$

or

$$e_B > \frac{0.0003}{0.0123} = 0.0244.$$

(Note that in the full program the shadow price of this constraint is $0.0257/lb.) Comparing vectors (2) and (3) in the same way shows that (3) will be preferred to (2) if $e_B >$ $0.040/lb. This is considerably lower than the shadow price of the 4,200-lb constraint in the full program ($0.0585), and reflects the very large "grid size" that has been created by choosing not to have any alternatives between the 7,200 and 4,200 constraints. A similar comparison for vectors (3) and (4) gives a critical effluent charge value of $0.100 from the smaller model versus $0.161 for the 1,800-lb limit in the larger model. If, however, there were only vectors (1) and (3), (3) would be preferred to (1) when $e_B >$ $0.034/lb; and if it were necessary to compare (1) and (4) directly the critical effluent charge would be $0.056/lb, much "worse" than the answer given by the four-vector model.

The lesson of this little exercise is that a condensed model will react to effluent charges in more or less the same manner as the full model; the finer the grid size of the computations used to set up the vectors of the condensed model, the closer the similarity of response. Similar examples could be developed for other residuals.

It is also true that, in general, if there are two vectors with the same discharge of a particular residual, one of which was developed using a constraint on that residual while the other resulted from constraints

TABLE 33. Illustrative Vectors for a Condensed Refinery Model

Row	No limits on residuals discharges (1)	7,200-lb limit on BOD (2)	4,200-lb limit on BOD (3)	1,800-lb limit on BOD (4)
Total crude charged (*bbl*)	1.0000	1.0000	1.0000	1.0000
Low-sulfur crude charged (*bbl*)	0.7400	0.7400	0.7400	0.7400
High-sulfur crude charged (*bbl*)	0.2600	0.2600	0.2600	0.2600
Premium produced, 100 octane/ $\leq$ 2.5 cc TEL (*bbl*)	0.2340	0.2340	0.2340	0.2340
Regular produced, 94 octane/ $\leq$ 2.5 cc TEL (*bbl*)	0.3410	0.3410	0.3410	0.3410
Waste heat (*10^6 Btu*)	0.3005	0.3005	0.3005	0.3005
NH_3 (*lb*)	0.0206	0.0188	0.0143	0.0075
BOD (*lb*)	0.0603	0.0480	0.0280	0.0120
Phenols (*lb*)	0.0321	0.0254	0.0125	0.0053
Sulfide (*lb*)	0.0028	0.0025	0.0017	0.0007
Oil (*lb*)	0.0470	0.0469	0.0467	0.0458
SO_2 (*lb*)	1.4289	1.4289	1.4289	1.4289
Particulates (*lb*)	0.4229	0.4243	0.4267	0.4286
Solids (*lb*)	0	0.0022	0.0062	0.0094
Objective function ($)	3.4813	3.4816	3.4824	3.4840

placed on some other residual, the former will be lower cost (or higher profit). Thus, for example, under the 4,200-lb BOD limit vector, the sulfide discharge is 0.0017 lb/bbl and the objective function value is $3.4824/bbl.

Setting a limit on sulfide discharge of 0.0016 lb/bbl and solving the full model results in a condensed vector having an objective function value of $3.4816/bbl, which would thus be preferred ceteris paribus to the vector with a 4,200-lb BOD limit as a way to react to a very high sulfide charge. Again a similar statement holds for vectors involving combinations of reductions. These comments, then, simply state that a condensed version of a full-scale model will be a better proxy the finer the grid of solutions is from which it is constructed, where this grid is defined not *only* on individual residual discharges, but also on combinations. This implies, of course, that in order to approximate the flexibility of the large model, a very large number of columns (i.e., alternative solutions) must be included in the condensed version. If there are five residuals of interest and if only a very rough grid (say high, medium, and low levels of discharge) is used, there are still 243 ($=3^5$) alternative solutions of the large model to be obtained and expressed in the appropriate vector form. If the grid fineness is increased to four levels of discharge, the number of solutions and vectors is increased to 1,024. Thus it is clear that the expense and bookkeeping involved in constructing collapsed models is considerable and that their column size can become very large.[11]

Whether or not the condensation approach could be a sufficient solution to the row-size problem depends, of course, on the number of rows in the resulting condensed models and the complexity of the region. If the average size of the individual models could be kept to ten rows, and the requirements for artificial bounds for the step-size selection part of the overall solution method are ignored,[12] a single linear programming model for a region of between 150 and 200 individual dischargers could be constructed. But keeping these condensed models to ten rows is not easy. Thus the inclusion of only one input, one output, six primary residuals, and two secondary residuals (sewage sludge and solids from particulate removal, for example), brings the model up to ten rows. Every refinement on the product or residuals side reduces the number of individual

[11] Column size is not a computational problem but does add to storage expense.
[12] Discussed in Clifford S. Russell and Walter O. Spofford, Jr., "A Quantitative Framework for Residuals Management Decisions," in Allen V. Kneese and Blair T. Bower, eds., *Environmental Quality Analysis: Theory and Method in the Social Sciences* (Johns Hopkins University Press for RFF, 1972).

sources that can be included. And the necessity for artificial bounds cannot be neglected, so that the "capacity" is very much lower than 150–200 plants; a guess would be 40–50. Now, for many regions this would be sufficient, but in a large industrialized region it would not begin to cover the significant point sources of airborne or waterborne residuals, particularly when it is realized that at least the largest municipal incinerators and sewage treatment plants must also be included and provided with discharge reduction alternatives.[13]

SOME FINAL NOTES ON FURTHER RESEARCH

While considerable work has been done, particularly for petroleum refining and the iron and steel industry, on the construction of models for scheduling operations, designing plants, and studying the industry in general, I am not aware of any work specifically directed to residuals generation and discharge within such models. Believing that models of the type described above are valuable, I suggest two obvious paths for further research.

First, the coverage of industries should be extended.[14] In particular, the chemical industries, major residuals sources wherever their plants occur, will require a very large amount of work. Organic, inorganic, and petrochemical plants may be approachable as groupings of individual

[13] The Philadelphia Air Quality Region Inventory of emission sources lists about 300 individual industrial plants, plus another 30 or so municipal incinerators and large institutional heating plants.

[14] Currently at RFF work is progressing on a model of an integrated iron and steel plant that builds on the work of Fabian, Trozzo Nelson, and many others and includes residuals generation and the many trade-offs implied by various process choices. See, for example, Tibor Fabian, "Blast Furnace Production Planning: A Linear Programming Example," *Management Science*, vol. 14, no. 2 (October 1967); Charles Louis Trozzo, *The Technical Efficiency of the Location of Integrated Blast Furnace Capacity* (Ph.D. dissertation, Department of Economics, Harvard University, 1966); and John P. Nelson, "An Interregional Recursive Programming Model of the U.S. Iron and Steel Industry: 1947–1967" (Ph.D. dissertation, Department of Economics, University of Wisconsin, 1970).

The RFF work is being done largely by William J. Vaughn. A study of the paper industry residuals situation by Blair Bower, George O. G. Löf, and M. W. Hearon is nearing completion. This study does not itself include a model, but the processes have been investigated in such meticulous detail and the residuals generation implied by alternative processes, output specifications, and so on, is so conveniently laid out that this next step should be relatively simple.

NSF-RANN is now funding research at the University of Houston, under Russell Thompson, which has as a major objective the building of models for chemical and petrochemical plants of particular types.

process units, just as the petroleum refinery is. If this is true, a logical approach is to begin by constructing alternative vectors (where alternatives can be identified) for these individual units. Then the task of constructing a model of a particular plant or plant type becomes largely a matter of combining these building blocks and providing the treatment, materials recovery, recirculation, and by-product production options available to the plant as a whole.

A second area for coverage extension is implied in the last sentence. Certain processes, particularly those for treatment, recirculation, and by-product recovery are applicable across a very wide range of industries. Consider, for example, standard activated sludge plants, electrostatic precipitators, and sulfur recovery plants. It would be extremely useful to have activity vectors for such processes, for varying sizes and performance rates, available "on the shelf," so that a researcher working on an industry would not have to construct these auxiliary units from scratch. It would also, of course, be useful if some effort, periodic or continuous, were made to study and provide activity vectors for new auxiliary processes. This should become more and more important as increasingly severe regulation encourages technical innovations that lower the cost of reducing discharges of the residuals considered most harmful externally. It is important, as I have stressed above, that models to assist in environmental quality decisions reflect reductions in costs of discharge control in order that there be no bias against a cleaner environment.

LIST OF ABBREVIATIONS

AC	Arabian mix crude
acf	actual cubic feet
acfm	actual cubic feet per minute
API	American Petroleum Institute
ASTM	American Society for Testing Materials
bbl	barrel(s)
BOD	biological oxygen demand
Btu	British thermal unit(s)
cc	cubic centimeter(s)
EPA	Environmental Protection Agency
ETC	East Texas crude
F.O.	fuel oil
FWPCA	Federal Water Pollution Control Administration
gal	gallon(s)
gm	gram(s)
ITT	International Telephone and Telegraph
KGO	coker gas oil
lb	pound(s)
LT	long ton
MGD	10^6 gallons per day
NAPCA	National Air Pollution Control Administration
NTIS	National Technical Information Service
NSF	National Science Foundation
ppm	parts per million
psi	pounds per square inch
psig	pounds per square inch gauge
RANN	Research Applied to National Needs
RFF	Resources for the Future
RON	Research Octane Number
scf	standard cubic feet
TEL	tetraethyl lead
VGO	virgin gas oil

INDEX

Acid, sprung: treatment, 72–73

Activated carbon absorption process: condensates treatment, 81–82

Activated sludge process: condensates treatment, 80–81; response to water-withdrawal cost, 130

Advanced refinery model: base solution, 118t, 119t, 120; BOD effluent charges, 141–42, 146; effluent emission charges, 173t, 174; particulate emission charges, 166, 168; phenol effluent charges, 148–49; processes and products, 59–61, 62t; sulfide effluent charges, 149–53; SO₂ emission charges, 161f, 162, 165

Airborne residuals: benchmark refinery, 101–5; from catalyst regeneration, 85t; direct influences on emissions, 157–69; discharge modifications, 82–86; emission charges, 173t, 175; from fresh heat, 87t; management costs studies, 7–8; from sulfur recovery, 75

Alkylation unit, 49

Alpert, S. B., 59n, 61n, 62t

American Paper and Pulp Association, 108n

American Paper Institute, 108n

American Petroleum Institute, 37n, 70n, 75n, 81n, 84n; petroleum refining study, 90, 92, 101n, 102t. *See also* Oil-water separator

American Society for Testing Materials, 51n, 55t

Ammonia: BOD reduction effects, 139, 140; effluent charges, 155, 173t, 174; refinery stream treatment, 74–82

Arabian crude oil. *See* Higher-sulfur crude oil

Arey, W. F., Jr., 53n

Aromatics, 44n

Ash, fly: disposal costs, 86. *See also* Particulates

Automobile catalytic converters: gasoline quality requirements, 110–11

Ayres, Robert U., 5n, 176n

Barnes, Thomas M., 8n

Basic refinery model, 35, 36t; benchmark solution, 89–106; BOD effluent charges, 138–41; particulate emission charges, 165–68; phenol effluent charges, 147–49; sulfide effluent charges, 149–53; SO₂ emission charges, 161f, 162f, 163, 165; technology effects on residuals generation, 109–10

Battelle Memorial Institute: airborne residuals costs studies, 8

Bechtel Corporation, 37n, 40t, 43n, 45t, 49n, 55t, 70n, 83n, 98n

Benchmark solution, 89–106

Beychok, M. R., 37n, 41n, 44n, 49n, 70n, 73n, 74n, 75n, 80n, 81n, 102t

Biological oxygen demand, 41n; effluent charges, 138–47, 154, 173t, 174; kerosene production increase effect, 178–79; refinery stream treatment, 74–82

Boilers: fractioning process use, 39; fuels burned, 84, 86, 87t; model application, 14–17; particulate control devices, 86, 88t

Bonner and Moore Associates, Inc., 35n, 111n

Bower, Blair T., 8n, 108n, 133n, 185n

Boyd, J. Hayden, 18n, 39n, 81n

Butanes: alkylation unit feed, 49; catalytic reforming, 43–44; catalytic cracking, 45, 46; isomerization unit feed, 43, 44. *See also* Isobutane

Burd, R. S., 81n

Cantrell, Aileen, 102n, 110n

Cardinal, P. J., Jr., 81n

Catalyst regeneration, 46–47; particulates treatment, 82–84, 85t

Catalysts. *See* Zeolite catalysts

Catalytic cracking: benchmark refinery, 96; coker gas oil, 48; hydrocracking technology effects, 109–10; virgin gas oil, 44–47

189

THE JOHNS HOPKINS UNIVERSITY PRESS

This book was composed in Modern No. 21 text and display type
by Monotype Composition Company. It was printed on Maple's 60-lb.
Danforth and bound in Holliston Roxite cloth by The Maple Press Company.

Library of Congress Cataloging in Publication Data

Russell, Clifford S
  Residuals management in industry.

  Includes bibliographical references.
  1. Petroleum refineries—Waste disposal. 2. Petroleum refineries—Mathematical models.
I. Resources for the Future. II. Title.
TD899.P4R87   665′.53   72–12367
ISBN 0–8018–1497–9